WHO OWNS THE WILDLIFE?

Recent Titles in
Contributions in Economics and Economic History
Series Editor: Robert Sobel

WHO OWNS THE WILDLIFE?

The Political Economy of Conservation in Nineteenth-Century America

James A. Tober

Contributions in Economics and Economic History, Number 37

GREENWOOD PRESS

WESTPORT, CONNECTICUT ● LONDON, ENGLAND

Library of Congress Cataloging in Publication Data

Tober, James A   1947-
   Who owns the wildlife?

   (Contributions in economics and economic history;
no. 37 ISSN 0084-9235)
      Bibliography: p.
      Includes index.
      1. Wildlife management—United States—History.
2. Wildlife management—Economic aspects—United
States—History.   3. Wildlife management—Law
and legislation—United States—History.   4. Wildlife conservation—United States—History.
5. Wildlife conservation—Economic aspects—United
States—History.   6. Wildlife conservation—Law
and legislation—United States—History.   I. Title.
SK361.T6   333.95'0973   80-23482
ISBN 0-313-22597-4 (lib. bdg.)

Library of Congress Catalog Card Number: 80-23482
ISBN: 0-313-22597-4
ISSN: 0084-9235

First published in 1981

Greenwood Press
A division of Congressional Information Service, Inc.
88 Post Road West, Westport, Connecticut 06881

Printed in the United States of America

10 9 8 7 6 5 4 3 2 1

People complain, and the legislature passes game laws, and nobody pays any attention to them after they are passed. Why? Because we insist on considering wild animals as our remote forefathers considered them, when men were scarce and wild animals were plenty. In a new country, the first settlers may properly have, not only liberty, but in some things license; license to till land anywhere, to cut wood anywhere, to shoot and trap game anywhere, to catch fish anywhere and in any way. All such things are then too plenty. As population increases, land and wood become PROPERTY. . . . This is the march of civilization; . . .

—Report of the Commissioners
of Fisheries of Massachusetts, 1868.

CONTENTS

TABLES AND MAPS

PREFACE

The snail darter is practically a household word, not because the three-inch fish offers hope to the protein-starved Third World, nor because it rivals the striped bass in its challenge to sport fishermen, nor even because it is the only known source of an enzyme that may offer relief to the sufferers of the common cold. The snail darter is well known because it is rare and because its only known habitat was first threatened and then destroyed by the completion of the Tennessee Valley Authority's (TVA) Tellico Dam on the Little Tennessee River.

Rareness has always engendered curiosity and often competition for limited examples of particular life forms, but only recently has it been accorded an explicit place in law. The 1973 Endangered Species Act (replacing the acts of 1966 and 1969) protects properly listed species of plants and animals from, among other dangers, federal government destruction of their "critical habitats." The snail darter thus halted the completion of the Tellico Dam and set in motion a congressional review that led to the 1978 amendments to the act. Among the most important changes was the creation of the cabinet-level Endangered Species Committee authorized to exempt federal agencies from compliance with certain provisions of the act.

The committee, finding that the costs of the Tellico Dam clearly outweighed its benefits, and unable to conclude that no reasonable and prudent alternatives existed, voted unanimously to deny the project an exemption. That finding invited Congress to legislate a specific exemption for the dam. Unless recent efforts to transplant the fish are successful, the snail darter will join the ranks of the extinct after all.

The protection of endangered species is but one critical issue in contemporary wildlife policy. Others include the allocation of funding between game and nongame animals, the management of wild horses and burros on public lands, the taking of porpoise in the tuna fishery, the regulation of international trade in animals and animal products, and predator and pest control.

The emergence of these issues is explained largely by the same forces that explain the environmental movement of the past fifteen years: affluence, leading to an expanded time horizon and a growth in the demand for environmental amenities; a total conversion of natural environments, increasing the value of remaining open space and its natural animal populations; and new technologies, complicating vastly an assessment of the costs of further growth. But the specific articulation of wildlife policy issues depends critically on the unique features of wildlife and on the manner in which these features have, in the past, constrained the policy process. Thus, the fugitive nature of the resource limits individual property rights in wild animals and links wildlife to virtually every decision in natural resource management. At the same time, the frontier heritage, which combines a high dependence on abundant wildlife with an escape from the restrictions of Old World institutions, continues to promote egalitarian access for consumptive uses of wildlife. But this is questioned by the ethics of control over other living beings. Intertwined with these strands is the complicated legal history of wildlife, which supports the proprietary interest of the states and limits federal jurisdiction.

Satisfactory understanding of current conflicts in wildlife management and their possible resolutions clearly recommends historical inquiry. This inquiry is well underway. The recent

works of James Trefethen, John F. Reiger, Theodore Whaley
Cart, Roderick Nash, Thomas A. Lund, and Michael J. Bean
speak variously to the role of sportsmen in nineteenth-century
resource management, the intellectual, scientific, and cultural
background to the contemporary conservation movement,
and the evolution of colonial and federal wildlife law. The
present work complements these with an analysis of state
regulation of access to wildlife during the nineteenth century.
The control which the states acquired, and the form in which
that control has been institutionalized, remain critical features
in the wildlife policy landscape.

Nineteenth-century journals and documents, and perhaps
especially the sporting literature, offer the twentieth-century
reader entertaining details, curious facts, and quaint phrases
that seductively undermine scholarly efforts to discern the
features of the larger picture. I gratefully acknowledge my
debt to those who have helped me to see these features. I am
particularly indebted to William N. Parker, without whose
long-term encouragement and helpful criticism this work may
not have emerged in print. Others who read portions of earlier
drafts include Clark S. Binkley, Roland W. Boyden, Colin W.
Clark, Richard L. Cole, Paul J. DiMaggio, Peter D. Hall,
George A. Hay, Richard M. Judd, Stephen R. Kellert, Michael
Krashinsky, Peter Kemper, John W. MacArthur, Joseph Mazur,
Robert C. Mitchell, Richard R. Nelson, Merton J. Peck, and
Sarah Sherman. Joann Nichols and Maureen Little at Marlboro,
and Andrea Compton at Yale, shared the burden of translating
edited copy into typescript. The preparation of an early draft
was aided by a Shell Assist Grant, awarded through Marlboro
College. The preparation of the present draft was greatly facilitated by the support and good fellowship of the Program on
Non-Profit Organizations, Institution for Social and Policy
Studies, Yale University. I owe my deepest appreciation to my
wife, Felicia, who aided me in numerous and onerous editorial
tasks, who willingly read and substantially improved several
drafts of the manuscript, who reminded me, in times of self-
doubt, that this book was worth writing, and who often knew
the word on the tip of my tongue.

INTRODUCTION

Before 1850, wildlife was more or less a free good. Residents of town and country alike hunted for food and skins, and they protected their property from predators and pests. The small but growing number of sport hunters found an abundance and a variety of targets. The market supported a vigorous national and international trade in furs and skins along with local trades in other wildlife products. There were a few state laws that offered protection to some species by closing the hunt during the breeding season, but the laws were infrequently, if ever, enforced. Whatever reductions in wildlife occurred through this unconstrained access were viewed largely as desirable and necessary consequences of economic development and growth. There was neither need for nor virtue in preservation.

This was to change, however, during the second half of the nineteenth century. Members of several influential and over-lapping segments of the population were able to articulate a broad and growing concern that the reduction in the abun-dance and distribution of wild animals was occurring too quickly. These sportsmen, naturalists, and humanitarians argued that even if wild animals provided no benefits other than meat and skins, free access offered few incentives for socially responsible harvest. Indeed, the very influence of these groups implicitly recognized new forms of social responsibility. But their argument was grounded in the further idea that wildlife provided benefits that ultimately were more important—animals could be hunted for sport, offering essential recreation to a population increasingly distanced from the rural environment, they could be studied as curiosities or as parts of a complex world, and they simply could be protected as monuments to the integrity of natural systems. Yet none of these demands could be satisfied without guaranteeing the maintenance of healthy, wild populations of animals. The decline simply had to be slowed, and this could be accomplished only by changing the conditions under which human and animal populations interacted—in effect by changing the structure of property rights in wild animals so as to prohibit or make less profitable certain further reductions.

This study focuses on the evolution of structured property rights in wild animals. It examines the articulation of conflicting demands for wildlife, the changes in animal populations which triggered them, and the character of the institutional responses to scarcity. Ultimately, it provides insights into the policymaking process, not the least of which is that the process is complex. The outcome is rarely the planner's outcome. It is complicated by a variety of institutional and cultural constraints; information is scarce; and the very essence of the problem comes to light only as the process plays itself out. The following paragraphs sketch the complexity of the process analyzed herein.

Among the general solutions available to the problem of resource management is the assignment of property rights to individual decision makers who, in promoting their own best interests, would husband the resources in society's best interest. Landowners are the parties to whom property in terrestrial wildlife might most logically be assigned. However, identifiable populations of most important species are not naturally confined by the boundaries of private landholdings. Thus, the benefits of management strategies undertaken by owners at their private expense will be shared with those over whose lands the animals roam. Domestication or artificial restraint such as fencing will resolve this difficulty only if the species in question can be successfully reared under restraint and if confinement does not destroy essential features of the populations. Given the patterns of land ownership that emerged in the United States, confinement would have lessened opportunities to hunt or observe many species "in the wild." The efficiency of private ownership is further limited by a variety of demands not readily translated into the vocabulary of the marketplace. Not the least of these is brought to light by the irreversibility associated with extinction.

The assignment of private rights would have posed a further, and equally serious, difficulty in threatening the customary freedom of hunters. Nondiscriminatory access to game animals had always been regarded as an important American liberty. The force of this notion effectively blocked the move toward private ownership without regard for the efficiency arguments.

The second general solution to the problem of resource management, and that which was to form the basis for the American approach, is the assignment of property rights to an omniscient administrator, generally an agency of the state. Such an administrator could, in theory, determine an optimal harvest for each species and regulate access in such a manner as to achieve that harvest. The process could be structured to maintain the equality of access threatened by the assignment of private rights. The ground for the move in this direction had been prepared by the tradition of the English game law and by the (perhaps erroneous) transfer of ownership in wild animals from the Crown to the colonies and then to the states. But the development of the game laws was hardly so orderly. Interest groups sought to limit the harvest in ways which threatened competing interests, states and regions worked at cross-purposes, and compliance was difficult to secure. Although the general regulatory structure derived from English experience, its particular content depended heavily on cultural heritage, on the distribution of political power, on the disposition of the courts, and on the configuration of wildlife populations.

The success of this solution was further limited by the primitive state of knowledge about the dynamics of wildlife populations in general and the role of habitat in particular. Although some animal populations were significantly diminished through hunting alone, the general decline in wildlife throughout this period can be attributed to the massive land-use changes that accompanied growth and development. Whatever appreciation there may have been for these dynamics, attempts to control the use of private land in the interest of wild animals would have been blocked by the strength of the sanctity of private property in land. There was to be little threat to the assumed right of landowners to use land in ways privately most profitable.

By 1900, the general dimensions of the resolution had been drawn. States had established a strong proprietary interest in wild animals and had enacted a variety of laws defining the seasons of the hunt and of sale, proscribing certain technologies, setting bag limits, and regulating access through licensing. These regulations were enforced by state-level agencies which gradually assumed advocacy and management roles. In that same year, Congress passed the Lacey Act, which introduced a significant federal presence in the management of wildlife. In addition to regulating the importation of species from abroad, the act gave effect to a variety of state laws which sought to regulate the passage of wild animals among the states. Thus, state regulation had, at the same time, come into its own and revealed its weaknesses.

It should be clear from this brief discussion that the particular evolution of property rights in wild animals depended critically on the cultural and institutional setting as well as on the theoretical concerns of resource management. Accordingly, much of this study seeks to convey a broad context for this process of change. Chapter 1 outlines the status of wildlife prior to 1850. It considers the early cultural and institutional responses to wildlife abundance and outlines the framework within which response to scarcity would occur during the period 1850-1900. During this latter period, there were three groups with a major stake in the outcome of the debate over the definition of property rights in wildlife—sportsman, market hunters and game dealers, and landowners. Chapter 2 describes these groups in terms of their stakes in this debate and of the strategies employed by their members in confronting scarcity. The behavior of these groups is not, however, independent of the scarcity which resulted from their own interactions with wildlife. Chapter 3 describes the changes in the abundance and distribution of wildlife populations during this period. It considers in turn, the effects of sport hunting, of market hunting, and of land-use change.

In the face of reduced abundance of wildlife, these three interest groups engaged to varying degrees in both market and nonmarket activities. Chapter 4 considers the market solution to scarcity and its institutional limitations. The final chapters examine the nonmarket solution. Chapter 5 outlines the scope of wildlife legislation during the period of this study and assesses the constitutional issues raised in the regulatory process. Finally, Chapter 6 considers the growth in state regulation of wildlife as viewed through several specific conflicts and their resolution. Appendix 1 more closely analyzes the structure of the three interest groups described in Chapters 1 and 2, and points out the implications of this structure for collective action. Appendix 2 provides a simple analytical framework for the analysis of some of the important dynamics described in the text.

WHO OWNS THE WILDLIFE?

1

VISIONS OF ABUNDANCE

There are Geese of three sorts vize brant Geese, which are pide, and white Geese which are bigger, and gray Geese which are as bigg and bigger than the tame Geese of England, . . . wild or tame, yet the purity of the aire is such, that the biggest is accompted but an indifferent meale for a couple of men. There is of them great abundance. I have had often 1000. before the mouth of my gunne, I never saw any in England for my part so fatt, as I have killed, there in those parts, the fethers of them makes a bedd, softer than any down bed that I have lyen on: and is there a very good commodity, the fethers of the Geese that I have killed in a short time, have paid for all the powther and shott, I have spent in a yeare, and I have fed my doggs with as fat Geese there, as I have euer fed upon my selfe in England.

Thomas Morton, *New English Canaan*, 1632

[This is] a hell of a place to lose a cow.

Ebenezer Bryce, as he first viewed the canyon which was to bear his name

WILDLIFE ON THE LAND

Hardly a single account of settlement or exploration in the New World failed to pay special attention to the variety and abundance of wild animals. In Georgia, the "woods abound with deer, and the trees with swarms of bees and singing birds. . . . There is great plenty of wild fowl . . . [a]nd the greatest variety of fish in the world."[1] In Virginia, two hundred deer "in one herd have been usually observed."[2] In Massachusetts, "I have seen a flight of Pidgeons . . . that to my thinking had neither beginning nor ending, length or breadth, and so thick I could see no Sun."[3]

These early accounts made good reading in Europe. The abundance of wild animals and the freedom to hunt them for personal gain were important symbols of liberty in the pictures painted for the purpose of generating settlement and financial backing for colonial ventures. Glowing reports might further serve to erase any doubts as to the wisdom of those who had already settled.[4] Where else but in the New World could civil servants receive payment in furs and skins,[5] and where else were game birds so abundant that "laboring people or servants stipulated with their employers not to have the [now extinct heath hen] brought to the table oftener than a few times in the week."[6]

These visions of abundance were repeated for other regions at later dates. The frontier, as it moved from the dense hardwood forests of the east to the prairies and the mountains, offered a succession of new environments, each uninhabited as compared to the one before it. Military surveys, naturalists' and travelers' accounts, emigrants' guides, land offerings, and railroad promotionals presented a steady stream of abundant visions not only of fish and game but of vast, natural wealth apparently free for the taking. In Kentucky, ornithologist Alexander Wilson observed a flock of passenger pigeons that he calculated to contain 2,230,272,000 birds, "an almost inconceivable multitude, and yet probably far below the actual amount,"[7] and Richard Irving Dodge estimated that 12,000,000 bison peacefully grazed the 1,250 square miles of plains within his gaze.[8]

In this literature, too, are the self-proclaimed testimonials to individual hunting prowess, an example of which is the autobiography of Meshach Browning attesting to a personal harvest of perhaps 2,000 deer, 400 bear, and 50 panthers. But especially this latter kind of record is of dubious generality. Frontiersmen who chose to record their experiences did so in part because of their extraordinary contact with wildlife. And wildlife was not uniformly distributed. Settlers "found wildlife everywhere because they traveled and settled where it was most concentrated"[9] and because the transitional vegetation that they created by clearing and other rudimentary improvements provided habitat for desirable species.[10]

Occasionally a more restrained vision survives, as for example that of the English settler in Wisconsin who, in the 1840s, wrote to his parents that

> some people form expectations of this cuntry before they come which would be impossible to realize in any cuntry in the world for I have thought sometimes that some people imagine that when they get to this cuntry they will find fish in every pool of water fruit on every tree and that wild fowl will come to them to be shot. . . I must Say there is plenty of Fruit and fish and fowl but they are the same in this cuntry as in any other, *no catch no have*. . . .[11]

These sobering visions notwithstanding, the early reports of abundance must have been striking in comparison to the resources of Europe, which had been settled for centuries. Yet, the personal hardships, the deaths, and the failures of whole communities attested to the insufficiency of wildlife, however plentiful, in guaranteeing survival in the New World. Because the European settlers' contact with wildlife was both deliberate and precipitous, they could not depend on patterns of interaction with wildlife such as those the Indians had established through centuries of mutual adaptation.[12]

THE CULTURAL LANDSCAPE

It is not wildlife abundance alone, or even abundance combined with the inexperience of new Americans, that explains the patterns of harvest and control that characterized early interactions with wildlife. Settlers brought with them ideas about natural environments. The New World was a wilderness and, for the Puritans at least, the wilderness was but a stage "through which we are passing to the Promised Land."[13] This promised land was something of a middle ground—a "city on the hill" or a "garden in the wilderness." In the middle and southern colonies, the wilderness was less harsh, but the reality of survival was no less pervasive. It was still a garden, though perhaps a grander one, with finer landscaping, which was to be carved from the wilds.[14]

Migration and settlement in the New World was motivated variously by religious, social, and economic considerations, and motivation set at least the initial style of the community. Of those who came to stay, the Puritans left perhaps the greatest legacy. But although their motivations for settlement were religious, the more mundane problem of survival attracted earliest attention. Wresting a livelihood from the harsh environment required a closely coordinated productive effort combined with the rapid acquisition of skills from neighboring Indians.

Were idleness not already proscribed by religious tenet, it would have been discouraged by these immediate economic considerations. Practically every diversion imaginable was initially condemned as a "carnal delight." Symbolic of the despicable life was Thomas Morton's "den of iniquite" at Merry Mount, raided in 1628 by a force of militia led by Miles Standish.[15] Recreation, in its root meaning of creating anew, was an essential ingredient in the productive, godly life. But without renewal, the same activities were mere frivolities. Thus, the Massachusetts General Court rebuked only "unprofitable fowlers," and Governor John Winthrop ("I have gotten sometimes a verye little but most commonly nothinge at all towards my cost and labour.") gave up fowling in response.[16] Even in non-Puritan New York, horse racing was permitted "not so much for the divertisement of the Youth alone but for the Encouragement of the bettering of the breed of horses which through great neglect is so Impaired that they afford very inconsiderable Rates. . . ."[17]

Hunting and fishing rapidly assumed important roles in the local economy, and fishing was to become a major source of foreign exchange for the New England colonies. But Puritan doubts occasionally clouded this economic vision. The Reverend Joseph Seccomb dismissed lingering concerns before his congregation of New Hampshire fisherman in 1739. To the charge "He that delighteth to see a Brute die, would soon take great Pleasure in the Death of a Man," came the reply, "We are so far from delighting to see our Fellow-Creatures die, that we hardly think whether they live." Since there is no doubt that man is permitted the consumption of fish,

the mere catching of them is no Barbarity. Besides, God seems to have carv'd out the Globe on purpose for a universal Supply . . . and he has implanted in Several Sorts of Fish, a strong Instinct to swim up these Rivers a vast Distance from the Sea. And is it not remarkable, that Rivers most incumbered with Falls, are even more full of Fish than others.[18]

Such strange behavior could have no other purpose than to provide food for man.

If the Puritan ideal could ever be realized, it was only in the small, organic village in which community norms were easily enforced and a unity of thought pervaded daily life. Church and State were not easily separated, and education fostered their union. But the forces which led to the dissolution of the community were built into its very fiber.[19] Fertile lands away from the village attracted the landless and those wishing to escape the watchful eye of community leaders. Increased distances promoted the settlement of new villages. In fishing lay great reward, and new communities were founded along the sea.[20] Some left in response to the perceived dilution of purity which accompanied growth, development, and continued immigration. The settlement of the Connecticut and New Haven colonies was motivated by these considerations.

As the perimeter of settlement grew, the new inhabitants confronted an increasingly large wilderness frontier and the problems of survival in a hostile environment. Wild country was simply inefficient in channeling the flow of energy through man. It stood in the way of settlement, growth, and the accumulation of wealth.

When our fore-fathers landed at Plymouth, what did they find, and what did they bring? Like the Roman Legions under Caesar, when he entered Britain, they found acorns, berries, and wild beasts, and men not less wild: But like Caesar, bearing to conquered countries the improvements that civilized his own, they introduced agricultural knowledge and skill, cattle and exotics, and lo! the wilderness "budded and blossomed like a rose."[21]

Just as it was for the landscape, so it was for the Indian. The Noble Savage of the French Enlightenment could not survive the true test of coexistence. European civilization destroyed the "simplicity" of Indian cultures, but, even in its purest form, this simplicity was significant only as a contrast to the cultured life, not as a substitute. More importantly, however, the Indians required land and resources that White settlers were unwilling to allow them. "The hunter or savage requires a greater extent of territory to sustain it than is compatible with the progress and just claims of civilized life ... and must yield to it."[22]

This view of the Indian was readily transferred to frontiersmen, social outcasts unable to make good in a society of hard work and a settled existence. Men such as this were, to Crèvecoeur, "no better than carnivorous animals of a superior rank. . . . By living in or near the wood, their actions are regulated by the wildness of the neighborhood ... once hunters, farewell to the plough. The chase renders them ferocious, gloomy and unsociable. . . ."[23] The fullest development of the breed was encouraged by the opportunities arising out of the fur trade in the Far West during the early nineteenth century. One student of western history describes the savage existence of the "free trappers" or "Mountain Men" as an illustration of the "extent to which environment transcended hereditary factors in molding the lives of the frontiersmen." The attitude toward the wilderness which arose out of the period of settlement is strong indeed. While few Mountain Men reverted to complete savagery as did "Cannibal Phil," who is reputed to have subsisted on human flesh during more than one hard winter, "all slipped far backward from a state of civilization." When it came to protecting their material interests, "*not even the savages of the forest could rival* [the Mountain Men's] cold-blooded cruelty and their callous indifference to human life."[24]

Wild animals were seen as largely responsible for the continued ability of the "uncivilized" to survive at a distance from the settled world. This view had major implications for Indian policy in the latter part of the nineteenth century, and was prominently exhibited in regard to the relationship between the Plains Indians and the buffalo, but the following passage from *A Report on the Quadrupeds of Massachusetts*, compiled in 1840 by Ebenezer Emmons at the order of the state legislature, suggests that it was not directed solely at the Indian.

So far as game and hunting are concerned, the sooner our wild animals are extinct the better, for they serve to support a few individuals just on the borders of a savage state, whose labors in the family of man are more injurious than beneficial. It is not, therefore, so much to be regretted that our larger animals of the chase have disappeared. What comforts their fur and their skins have provided, can be abundantly supplied by animals already domesticated, at far less expense, both of time and money, and are not subject to that drawback, the deterioration of morals.[25]

The rejection of the wilderness did not imply a desired state in which the last vestiges of the natural world had been covered over. What had been the Puritan "city on the hill" became the agrarian republic of the eighteenth and early nineteenth centuries. The middle state, no longer an option for England, could be a reality in America with its sprawling, unclaimed lands always to the west, its egalitarian institutions, and its young, vital population. For Thomas Jefferson, the choice between rural virtue and economic growth was easily made. Allow Europe to industrialize, and trade for what was needed. "The loss by the transportation of commodities across the Atlantic will be made up in happiness and permanence of government."[26]

But the ideal perhaps did not structure social and economic life so much as it emerged out of that life. Amid great natural abundance, the absence of constraints on resource use encour-

aged behavior that coincided with agrarianism, serving to for-
tify the myth and give it time to simmer and develop in the
minds of those few who would take it upon themselves to
define a national character for a nation which, by virtue of its
newness, had none.

Not only may the myth have derived from economic realities
rather than the reverse, but the extent to which the two were
coincident may be questioned. There were those in the commercial and nascent industrial spheres who were never even
close to the ideal, and who likely felt neither sinful nor sorry
about their materialistic endeavors. In fact, there were likely
few farmers, except perhaps those who returned, independently
wealthy, to the land after a career in politics or law who were at
all compulsive about retaining the self-sufficient nature of
agriculture merely out of respect for rural virtue. By 1816, even
Jefferson had changed his mind. He who now opposes domestic manufacture must favor reducing the nation to dependence on foreign powers "or to be clothed in skins, and to
live like wild beasts in dens and caverns. I am not one of
these. . . ."[27] The recent war had demonstrated the consequences
of dependence and had offered great incentive to domestic
manufacture.

While agrarianism gave way to nascent industrialism, pas-
toralism did not. The American environment would purify
industry, leaving only the desired effects of increased output
and leisure. It was the difference between oppression by the
factory system and liberation by the machine. "One thing is
plain for all men of common sense and common conscience,"
wrote Ralph Waldo Emerson in "The Young American," "that
here, here in America, is the home of man."[28]

Sheer abundance and a persistent faith in progress promoted
this perception. Thomas Ewbank, commissioner of patents
from 1849 to 1852, expressed the dominent philosophy in *The
World a Workshop, or the Physical Relationship of Man to the Earth,*
published in 1855. Here, he developed the theme that the
world's resources were inexhaustible and that the earth was
designed by the Creator to be worked by man following the

principles of science. "It is preposterous to suppose that the supplies of coal can ever be exhausted or even become scarce. The idea is almost blasphemous."[29] And if some were forced to concede a limited supply of coal, they were, at the same time, certain of its replacement by peat just as coal had replaced wood.[30] The land itself was so abundant that "a period of five centuries must elapse before the whole public domain of the Union . . . will become the property of individuals."[31]

The middle state was promoted both by the advocates of agrarianism and the advocates of industrialism. Even the railroad, ultimately the symbol of the conflict between these paths, initially brought them together by promoting a national division of labor and its consequent prosperity. But as the railroad and industry grew, the sanctity of the small landholder was threatened, and, ultimately, industrialization, in the name of freedom and equality, swallowed up the freeholder, in the name of progress and growth.

THE NATURAL LANDSCAPE

For a growing number of spokesmen, America's uniqueness lay not in her ability to transform the wilderness into a garden and to harness the machine, but in the wilderness itself, which, "by the middle decades of the nineteenth century . . . was recognized as a cultural and moral resource and a basis for national self-esteem."[32] But European observers were unwilling to accede to these claims. The most outspoken critic, Comte de Buffon, author of the forty-two-volume *Histoire Naturelle, Générale et Particulière*, sought evidence in America for his grand theory that the New World had been created after the Old, by which time the "basic molecules of life" had lost their vitality. "The place can afford nourishment only for cold men and feeble animals."[33] Some European rulers uncritically promoted reports of this kind to offset the visions of abundance which contributed to the emigration of their subjects.[34] The claims that New World species were but degenerate forms of Old World life were, of course, patently false, but not so obviously so that they did not merit serious attention by the likes of Benjamin Franklin, Thomas Paine, John Adams, and Thomas Jefferson. Jefferson's *Notes on the State of Virginia* was in part a direct response to the charges.[35]

The debate was largely diffused by increased communication and travel and the growth of science. The quality and uniqueness of the American environment were confirmed by generations of natural historians, explorers, and migrants who catalogued and collected indigenous species and recorded their impressions of the natural world, and by increasing numbers of sportsmen and tourists who followed in their paths. Natural history cabinets, housing curiosities of nature, become museums, offering organized collections of specimens from all realms of the natural world for public edification.[36] Natural histories, such as those of Alexander Wilson and John James Audubon, finely illustrated and expensively produced, were sold by subscription to underwrite the costs of their multi-year publications. Audubon searched two continents for subscribers to his monumental *Birds of America*, and, over the course of production, lost and gained them regularly as fashion in natural history changed. During one period of decline in English support, Audubon observed that "The taste is passing for Birds like a flitting shadow—Insects, reptiles and fishes are now the rage and these fly, swim or crawl on pages innumerable in every Bookseller's window."[37] Changes in printing technologies made even fine editions more affordable, but, meanwhile, modest editions that systematically explored broad classes of species grew in popularity.[38]

These developments which brought the natural world to the people were complemented by a preservationist movement, begun primarily by a small cadre of eastern, urban, elite visionaries who sought to protect wild places as repositories of unique natural features and as sanctuaries from a hectic world. First George Catlin, and later Henry David Thoreau, envisioned remaining portions of the American wilderness "containing man and beast" preserved "not for idle sport or food, but for inspiration and our own true recreation."³⁹ Beginning with George Perkins Marsh's remarkable *Man and Nature*, published in 1864, the movement gradually acquired a scientific perspective that led to more utilitarian arguments for preservation, for example, in the protection of watersheds for flood control and drinking water supplies.

The vitality of preservationism was stimulated by rapid changes in the landscape. Forested lands were cleared for farms, and farms were abandoned to the forest. Towns grew into commercial and industrial centers, and new settlements sprang up along rivers and along an expanding network of roads, rails, and canals. Changes in land use brought changes in wildlife abundance and diversity. Wild species differ in the extent to which they prosper in the environments created by man. The large carnivores and the species which require an unbroken expanse of uniform habitat, such as the passenger pigeon and the bison, are the first to be affected by change. Other species, such as the deer and the quail, do well in regions sparsely settled by human populations. Clearing creates favorable edge environments, and agriculture brings choice vegetation and insect food. As edge environments become more scarce, these species decline. With increased density of human settlement and accompanying habitat change, only domesticated and pest species remain.

THE EARLY HARVEST

The changes in the distribution and abundance of wildlife populations occasioned by alterations in the landscape were accentuated, from earliest settlement, by the extensive hunt for meat and skins. The demand for meat was largely satisfied by subsistence hunting but, as cities grew in the east, commercial or "market" hunting developed to supply nearby urban markets. The extent of the trade was limited by spoilage in transit and by the relatively small segment of the population which was both able to afford game and unable to find convenient access to the hunt.

The two most important commercial trades were in deer skins and furs. While the deerskin trade existed primarily in the southern colonies, Thomas Morton recognized its profitability in New England in 1632.[40] The trade was sufficiently brisk in Rhode Island that, in 1638, the colony controlled prices and named liaisons with Indian traders.[41] In 1677, the export of deer skins, "which are so serviceable and usefull for cloathing," was prohibited in Connecticut under penalty of forfeiture of the merchandise. Two years later, constables "who are water bayleys" were ordered to search vessels for illegal exports. The trade was opened again in 1681 but with the stipulation that an exporter had to first accept any domestic offer of six pence per pound or more.[42]

The southern trade was centered in Charles Town. Colonial agents ranged far into Indian territory, bartering western goods for the products of Indian hunting skills. Before rice became the staple crop, this trade was the basis of the export economy. As late as the mid-eighteenth century, skins brought greater export earnings than indigo, cattle, pork, lumber, and naval stores combined.[43] Deer remained an important item in both domestic and foreign trade long after the trade diffused from Charles Town. The huge volume of skins was accompanied by a great supply of venison which was available in major city markets "at so cheap a rate as to bring it within the means of almost every housekeeper."[44]

Of the fur-bearing mammals, the beaver was the most important, though the muskrat, mink, weasel, raccoon, and otter did not escape human predation. The beaver trade was opened in the New World by Jacques Cartier, who shipped abundant furs, which had been gotten from the Indians of the St. Lawrence region, to France. During the next two centuries, as European influence moved west and south, the French, Dutch, and British competed for alliances with the Indians and for control of the fur trade.[45]

American interest in fur trade did not emerge until the nineteenth century. Before that time the Revolution and the problems of the new nation demanded immediate attention. Because there was no significant American activity on the frontier, the Americans had little influence with neighboring Indian tribes, whose friendship had been cultivated by the Spanish and British. The Indian attacks on American frontier settlements were frequent, so frequent that in 1795 the federal government appropriated a small sum to establish trading posts in the hope of generating both friendly relations with the Indians and profits. The success of the trade, combined with increasing domestic demand for furs (much of which had been supplied by Canadian imports) and the acquisition of the Louisiana Territory, led to a rapid growth of private interest in the trade. John Jacob Astor's American Fur Company, chartered in 1808, was followed by the Missouri Fur Company in 1809 and in 1822 by the Rocky Mountain Fur Company. The intense competition that followed was ended about mid-century, largely the result of a change in fashion.[46]

Accompanying the market hunt was a small, but growing, interest in sport hunting. Although rural sports did have some advocates before the mid-nineteenth century, the character of the American approach in these early years was lacking the refinements which, according to continental tastes, differentiated the barbaric from the urbane. William Cobbett wrote, in 1819, that there "cannot be said to be anything here, which we, in England, call *hunting*. The deer are hunted by *dogs*, indeed, but the hunters do not *follow*. They are posted at their several stations to *shoot* the deer as he passes. This is only one remove from the Indian hunting." And if poor style were not sufficient indictment, the "general taste of the country is to *kill* things in order to have them to *eat*, which latter forms no part of the sportsman's object."[47]

Civility was not entirely lacking. Even in the Puritan north, genuine sporting interests developed as the frontier moved west, leaving behind cities and social classes. In 1783, "A Gentleman," in *The Sportsman's Companion*, described the upper-class amusement of bird hunting. The suggested provisions for a two-day hunt for a pair of gentlemen included "a single horse chair, a Servant in a second chair, to carry the

Dogs...provisions, liquors, tea, sugar, etc..." an outing which, no doubt, would have tempted Mr. Cobbett.[48] Even in cities to the west, in less than a generation from the time when every settler owned and regularly used firearms, shooting had become a novelty. Hunting clubs formed and formal hunts were undertaken.[49]

The hunts promoted social intercourse, more than sport, among participants, and competitive elements were prominent. In the side-hunt, long popular in both the east and the west, species were assigned points in relation to their desirability and scarcity, and competing teams went into the woods after the highest point total. Another common, if less refined, form was the drive or ring hunt in which participants would encircle a large area and gradually close in, driving ahead of them whatever game could not escape. "Guns would be used as long as this was reasonably safe, and then, clubs, pitchforks, any available weapon."[50] By the mid-nineteenth century, bull's eyes often replaced live targets, and marksmen competed for prizes. In one locality, where a beef animal was to be divided among competitors, custom decreed that to the first marksman went the hide and tallow of the animal, to the second the hindquarter of choice, to the third the remaining hindquarter, to the fourth and fifth the forequarters, and to the sixth the lead in the tree on which the targets had been affixed.[51]

It was merely, then, that refined sport was impractical and unpopular, not unknown.[52] To hunt for food and skins was one thing, but to sacrifice efficiency for style was quite another.[53] The New England aversion to recreation for its own sake persisted in some circles, in a form more extreme than in the Puritanism that inspired it, late into the nineteenth century. Hans Huth cites the Beecher Sisters' *Housekeepers Manual*, which in its 1873 edition, reminded the reader that "hunting and fishing, for mere sport, can never be justified."[54] But even absent the moral qualifications, one must always choose in life, and "I cannot but think that shooting and fishing in our state of society, must always be indulged at the expense of something better."[55]

PATTERNS OF CONTROL

The waves of settlement and accompanying commerce left nineteenth-century America with a very different complex of wildlife populations than had existed some two centuries earlier. Several species and races had become extinct, others gravely threatened, and most greatly reduced in their range.[56] In many cases, these scarcities were well noted as they developed. John Josselyn remarked in 1672 that "the English and the Indians having now destroyed the breed, . . . 'tis very rare to meet with a wild turkie in the woods."[57] Astor foresaw that the decline of his own industry would result from the "indiscriminate slaughter practiced by hunters, and by the appropriation to the use of man of those rivers and forests which have afforded [the beaver] food and protection."[58] Ornithologist J. A. Allen attributed the great decline in Massachusetts birds to "inevitable and natural causes" such as the removal of forests and the cultivation of land.[59]

But the realities of increasing scarcity could not overcome the pervasive idea that the sheer abundance of resources seemed to preclude any but local scarcities. The common view was that resources were simply inexhaustible. In 1857, the Ohio Senate quickly disposed of a bill which would have offered some protection to the passenger pigeon with the observation that the birds were so "wonderfully prolific" that "no ordinary destruction can lessen them, or be missed from the myriads that are yearly produced."[60] The destruction that followed was extraordinary. Habitat was destroyed by lumbering and settlement, and birds, dead or alive, were shipped by the millions to market. The passenger pigeon was scarce by 1890, and Martha, the last known member of the species, died in the Cincinnati Zoo in 1914.

Whether these described changes in wildlife abundance were local or widespread, whether they were necessary or fortuitous, whether they were desirable or tragic, it is clear that they were the result of activity that was almost without legal restriction. Wildlife was there for the taking. It was a common heritage, not subject to restrictive controls which smacked of

Old World class structure. In the American mind, equality meant equality of opportunity, not equality of wealth and power. The obvious disparities that emerged over the course of American growth and development were less a threat to equality than a confirmation of its existence. David Potter observes, in his classic essay on the American character, an ideology that supports not "soak the rich" but "deal me in" the lottery.[61] It is this image which so strongly protects fish and game from the royal control that befell it in Europe—the lottery here being a fair chance in the hunt.

In the early periods of settlement and exploration, this lottery applied equally well to all resources, and public policy was molded largely by the patterns of resource use which emerged from unconstrained behavior. Federal land policy recognized the claims of squatters who anticipated orderly settlement of the public domain by settling on choice parcels; mineral law legitimated and institutionalized the major patterns of interaction which sprang up on the mining frontier; the attempted enforcement of the federal government's property rights in standing timber on the public domain fell before the customary patterns of use on the frontier; and water law in the arid west grew out of patterns of use and tenets of Spanish law, which were more suited to the hydrology of the region than was the English doctrine of riparian rights established in the east and familiar to Washington policymakers.

There were, of course, extended debates in state and federal legislatures over the appropriate management of natural resources and numerous enactments limiting free access. Colonial regulations were severely limited by abundance. Enforcement was costly, and, were the costs met, the consequence may only have been migration to unmanaged areas. Nineteenth-century debates were usually resolved in favor of private ownership and the stewardship that was supposed to result. Federal lead and salt deposits, for example, were initially allocated by short-term leases, but enforcing the leases would require "the creation of a new corps of federal . . . agents." Even then the system would be "less advantageous to the Union than if the mines were committed to the care and ardor of individual enterprise."[62] And, indeed, leases were abandoned in favor of sale. Timber on the federal domain did not receive congressional attention until 1831, and for the next twenty years, the prohibition of cutting and removing timber was weakly enforced by the office of the solicitor of the Treasury. Responsibility was shifted to the new Department of the Interior in 1852, and agents were appointed in regions of greatest violation. But congressional critics saw only "paternalism inimical to the spirit of American freedom," and enforcement was effectively abandoned following the transfer of authority to the General Land Office and its local land officers.[63]

Wildlife, by its fugitive nature and its importance as an indicator of egalitarianism, remains largely distinct, and the regulations that structure its exploitation speak to its uniqueness. Other fugitive resources are oil and gas which, although fixed in place, can be extracted from common pools at multiple locations, and water, which although confined to waterways, moves through the holdings of multiple owners.[64] But wildlife, alone among natural resources, received constitutional attention. The "liberty to fowl and hunt upon the lands they hold, and all other lands therein not enclosed; and to fish, in all waters in the said lands," offered in William Penn's charter of 1683, was not an idle gesture to the inhabitants of Pennsylvania.[65] It was the articulation of deeply held beliefs about New World freedoms—beliefs that persist to the 1980s and which underlie much of the conflict in the present management of wildlife resources. The Pennsylvania Constitution of 1776 continued the liberty to fowl and hunt "in seasonable times" (in recognition of the already well-known need to protect breeding populations) and to fish "in all boatable waters, and others not private property."[66]

The following year, Vermont offered its citizens a slightly amended liberty to hunt and fowl and "to fish in all boatable and other waters (not private property) under proper regulations. . . . " The significance of the deviation from the Pennsylvania text from which it was derived came to light in the 1895 case of *New England Trout and Salmon Club v. George Mather*, which turned first on whether the parenthetical "not private property" modified "boatable" as well as "other waters" and second on whether "boatable" meant boatable in fact or navigable as defined by English common law. The majority, following the lead of other states in support of private economic growth, found for the plaintiff and upheld private property in inland waterways. Justice Laforest H.Thompson, in articulate dissent, took the occasion to review the merits of unrestricted access to fish and game. The Founding Fathers, he said,

were then smarting under the oppression and inequalities of the English system under which individual development among the common people was impeded and often prevented, and the rights and enjoyments of the many were subjected to the pleasure of a favored few. Among the instrumentalities used to bring about the undesirable condition of life, were the iniquitous fish and game laws of England, enacted by the ruling class for their own enjoyment, and which led to a system under which the catching of a fish or the killing of a rabbit was deemed of more consequence than the happiness, liberty or life of a human being. The framers of our constitution knew that the English system of fish and game laws had been a most fruitful source of crime and misery, and I think it was their purpose to so provide that this state should never be cursed by a like system. They believed that the raising of men was of more importance than the breeding of fish for sport or profit. . . .
This freedom to associate with and enjoy nature, has borne fruit in the independent, liberty loving character of our people, and has had its influence in forming a type of manhood that has had a potent influence in making Vermont to-day in many respects, the ideal republic of the world.[67]

The character of the wildlife and the cultural heritage which surrounds it cannot alone explain its institutional treatment. Broader developments in legal history merit consideration. American colonial law, as such, did not exist. The law in individual colonies depended on the stage of legal development at the colony's founding, on the character of the region, and on the character of its people.[68] Legal code rather than common law guided behavior, and that code, at least in New England, spoke to a wide variety of behaviors.[69] Highly significant was the fact-finding and law-finding power of juries and the consequent importance of community standards over either colonial statutes or English precedent. Thus, at least in Massachusetts, "the legal system could not serve as an instrument for the enforcement of coherent social policies . . . when those policies were unacceptable to the men who happened to be serving on the particular jury."[70] This feature persisted well into the period covered in this study, especially in the western states and territories, which were passing through their own "colonial" stages, and seriously hampered the enforcement of state game laws, the provisions of which were regularly discarded by juries of peers.[71]

Critical to the unraveling of a structure for the regulation of access to wildlife is the development of conceptions of property in land. In the colonial period, the variety of legal and customary forms that restricted the use and transfer of land (encumbrances and primogeniture) rapidly collapsed into the recorded land deed, simple and streamlined documents, and the transfer of title free and clear. The huge volume of transactions and the rapid turnover of individual parcels made this change both convenient and necessary.[72] Land, once the unique resource that integrated the town and separated it from the outside world, was now a commodity little different from other material goods, and landowners sought to exercise complete control over their domains. Ownership of wildlife on the land seemed to some spokesmen to follow logically from ownership of the land itself.

But as the clarity of title improved, the rights of property owners to the "quiet enjoyment" that protected their claim to an agrarian life was being eroded. Under the pressures of commercial and industrial development and under the conflict of increasingly dense settlement, the formerly compatible doctrine of prior appropriation became an aggressive strategy for promoting development by rewarding risk taking, "quiet enjoyment" notwithstanding. In turn, property rights acquired by prior appropriation were challenged under the "balancing test or 'reasonable use' doctrine which sought to define the extent to which newer forms of property might injure old with impunity."[73] The notion of compensated taking was not firmly accepted until about 1850, by which time, according to Morton

J. Horwitz, much of the redistribution favoring growth had already occurred.[74] Parallel developments in tort law saw the decay of the Blackstonian notion holding that all actions, whether or not lawful, were actionable and its replacement by 1840 with the "principle that one could not be held liable for socially useful activity exercised with due care."[75]

The evolution of law during the early national period was left largely to the states, which appropriated the common law of England. The post-Revolution Anglophobia, which led some states expressly to forbid reference to English cases or documents, offered only a brief break in the otherwise whole-sale adoption.[76] Yet, some conflicts made clear that English law was unsuited to the American environment of resource abundance and population scarcity. According to English law, for example, stolen wild animals, upon having been reduced to private property, were the subject of larceny only if they were fit for the food of man or met other specialized criteria, as did hawks or falcons. Otherwise, theft was only a misdemeanor. *State* v. *House* concerned the theft of an otter from a trap, an act judged by the lower court to be a misdemeanor. The North Carolina Supreme Court found the act larcenous, arguing that the "English system of game laws seems to have been estab-lished more for princely diversions than for use or profit, and is not at all suited to the wants of our enterprising trappers. ... We take the true criterion to be, the *value* of the animal, whether for the food of man, its fur or otherwise."[77]

As states and territories grew in number and as commerce among them supplanted self-sufficient local economies, the diversity of state legislation and the weakness of federal law became serious liabilities. The relatively short legal history in the United States limited the work of consolidation, and the development of legal thought based on precedent was more narrowly channeled by the American court practice of the majority opinion, which replaced the multiple opinions of English courts. Federal courts availed themselves of the oppor-tunity to expand federal jurisdiction in the name of economic growth, as for example, in the broadening of the admiralty clause to include inland as well as coastal waters.[78]

LAW FOR WILDLIFE

By English common law, animals, *ferae naturae*, were the property of no individual until reduced to possession by capture or control. But the colonies, and later the states, in promoting the general welfare of their citizens, limited the conditions under which individuals could acquire private rights in wildlife. Similarly, they encouraged the destruction of wildlife deemed to be a common threat. Implicit in this authority was the states' proprietary interest in wild animals, an interest that had been acquired by direct transfer of ownership from the English king to the colonies. The ownership doctrine, explicitly recognized by the courts in the last quarter of the nineteenth century, has served to channel twentieth-century federal wildlife policy through the enumerated powers expressed in the treaty-making, property, and commerce clauses of the Constitution.[79]

Colonial governments exercised their control over wildlife with great regularity, as may be demonstrated readily by an examination of the statute books.[80] Two types of controls predominate—those that presume to regulate the harvest of valuable species, notably deer and fish, to preserve a breeding stock that might otherwise be consumed by unregulated taking of the common resource; and those that encourage the destruction of predators and pests in recognition of the collective responsibility to provide a safe environment for the conduct of individual and community affairs. The development of Connecticut statutes protecting deer and persecuting wolves suggests the range of concerns.

Connecticut enacted its first wolf bounty in 1647. Ten shillings were awarded for each wolf killed within the boundaries of a town in the colony. The bounty remained in effect, with periodic changes in payment and eligibility, until 1774.[81] The history in other colonies and for other predators was similar. By the close of the colonial period, predators had been virtually eliminated from the settled areas of the east. Large sums of money were thus spent to encourage an activity that, to the extent that predators were a real threat to the settlers, ultimately would have occurred without compensation. Moreover, the bounty may have served to encourage people to enact clever schemes for collecting on counterfeit evidence and to induce them to hunt selectively to preserve a breeding stock and, thus, a future stream of bounty payments. But the bounty may also have encouraged an efficient division of labor, and it did promote killing the animals rather than merely devising schemes to reduce the impact of their depredations. This latter distinction was brought to light in 1717 when the several towns of Cape Cod planned to construct a wolf-proof fence between Wareham and Sandwich to provide safe pasturage on the outer cape. The inland towns defeated the measure with the argument that it would force upon them a larger share of wolves than was their obligation to accept.[82]

Although wolves had no recognized positive value in their wild state, captured wolves had value in the bounties they would bring. Connecticut recognized this value in 1656 by prohibiting the removal of any wolf from a pit made by another resident for the purpose of catching wolves if the intention was to defraud the owner of his rightful bounty. Another wrinkle in the bounty laws concerned the fear, expressed in 1716, that wolves killed outside of the boundaries of Connecticut were "by Indians or others brought in and sold for a small matter to persons destitute of an honest principle, living within this Colony, and by such frauds obtain bills drawn upon the treasury of said Colony." The Assembly ordered constables and selectmen, before making payment, to ascertain that wolves brought for payment were killed within the Colony.[83]

The colonial protection afforded deer seemed directly motivated by the decline in its population. In 1698,

Whereas the killing of a deer at unseasonable times of the year hath been found very much to the prejudice of the Colonie, great numbers of them having been hunted and destroyed in deep snowes when they are very poor and big with young, the flesh and skins of very little value, and the increase greatly hindered . . .

the Connecticut Assembly prohibited hunting until July 15, 1699, and thereafter from January 1 to July 15 of each year.

Successive offenses were to be met with increasingly higher fines. The possession of fresh meat or skins during the closed season was accepted as evidence of violation. Enforcement was to be encouraged by the division of the fine between the informer and the colony.

Without explanation, the law was repealed in its entirety in 1701, only to be reinstated, almost verbatim, in 1715. The new hunting season opened on July 1. In 1717, the fine schedule was altered to refer to the number of illegal deer taken rather than the number of offenses. In 1739, the closed season was extended to the end of July.[84]

Although the motivation for the first Connecticut law may have been in the perceived shortages of deer, the wording appears to derive from an identically phrased statute passed some four years earlier in Massachusetts. The length of the initial closed term, the length of subsequent hunting seasons, and the structure and disposition of fines were all identical. The wording was well received, for New Hampshire's first deer law, passed in 1740, was prefaced by the same remarks. By this time, however, both Massachusetts and Connecticut had extended their closed seasons to July 31, and so New Hampshire selected that date. Further, Massachusetts had, in 1739, required each township to appoint two deer-reeves to enforce the law, and New Hampshire required the same.[85]

The character of game legislation changed little during the colonial period. The deer was the most prominently mentioned species, though by 1708, New York law protected the turkey, heath hen, partridge, and quail in Suffolk, Kings, and Queens counties.[86] Later, the laws in some areas (Tennessee in 1741, Delaware in 1795, New Jersey in 1798, Alabama and Mississippi in 1803, Ohio in 1805, Indiana in 1807, and Illinois in 1821) prohibited Sunday hunting and the laws in some areas (Maryland in 1730, South Carolina in 1769, North Carolina in 1777, Georgia in 1790, Virginia in 1792, Alabama and Mississippi in 1803, and Florida in 1828) prohibited fire-hunting.[87]

The management of the inland fisheries offers a different history. Fish, just as terrestrial wildlife, could be reduced to private property only by capture, but, unlike wildlife, their habitat is well defined by the boundaries of watercourses. Free access to fisheries might well conflict with property rights of abutting landowners, but even were access limited to these owners, strategically placed nets or weirs would severely limit the catch of other owners. Overinvestment in equipment and overfishing might logically result from an open scramble to lay claim to the common property in fisheries. Efficient management suggested, in effect, the regulated monopoly in which protected access was granted or sold in exchange for controls over fishing technology, market prices, and distribution of product.[88]

But just as towns, in limiting access, had recognized the limitations of unconstrained harvest within their jurisdictions, so the colonies recognized the collective limitations of monopolies granted by several towns along the same watercourse. In 1766, Connecticut nullified the authority of local governments to permit the obstruction of waterways. This permission

is often found to prove mischievous, and many persons having obtained the same have thereby almost wholly stopped and obstructed the natural course of the fish, . . . and thereby have greatly destroyed the common privilege and benefit of taking and catching the fish.

The repeal of the statute under which many had received monopoly privileges led to a rash of requests for the continuance of special privilege. In each case, the state assembly assigned a committee to inspect the region and to "make such regulations for the carrying on of . . . [the fishery] . . . as they shall judge best for the public advantage," upon which the assembly would make a determination. In the case of Timothy Tiffany, the report was issued some two years after the initial complaint. It provided that Tiffany, located one hundred yards from the mouth of the Eight Mile River in Lyme, would be permitted to fish with a seine net Monday through Friday during daylight hours and only from his own property; that no one would be allowed to fish between his property and the mouth of the river; that all property owners above his location would be allowed to fish at their discretion; and that Jonathan Wade, located further up the river, would be allowed to erect a weir for use at ebb tide only.[89]

The problems that faced first towns and then colonies and states later faced groups of states that share access to fisheries. Several unsuccessful eighteenth-century attempts to secure cooperation within New England paved the way for limited successes following the creation of state fish commissions and the U. S. Fish Commission in the last third of the nineteenth century.[90]

A crosscutting trend, however, amplified claims to private property which suited the new regime of growth. In Massachusetts, while control over the fisheries of navigable waters remained with the towns, the definition of navigability was restricted from the broader "navigable in fact" to tidal waters. The Connecticut River fisheries, controlled by towns through the Revolutionary period, were thus converted to the private property of riparian owners.[91]

Colonial fish and game laws were selectively enforced at best. Helenette Silver, in her study of New Hampshire wildlife history, was unable to discover a single instance of the enforcement of the law protecting deer in that state prior to the 1878 creation of the Office of State Game Commissioner.[92] The New Hampshire legislation may have deterred some illegal behavior, but it was more often unenforced despite widespread violation. George Bird Grinnell suggested that early legislative efforts were "merely an inheritance from English ancestors" and never taken seriously.[93] Indeed, immigration to the American colonies was motivated by an effort to escape "the penalties of the harsh game laws and odious forest laws" of Europe.[94] Certainly the subsistence hunter felt justified in ignoring the letter of the law in supplying his larder in times of leanness.

The legislative attention to game laws fell off dramatically during the early period of statehood. Contemporary observers noted few restraints. A visiting Englishman found in 1819, that

as to *game laws* there are none, except those which appoint the *times* for killing. People go where they like, and, as to wild animals, shoot what they like. There is the Common Law, which forbids *trespass*, and the Statute Law, I believe of *"malicious trespass,"* or trespass *after warning*. And these are more than enough; for nobody that I ever hear of, *warns people off*.[95]

Several years later, "A Gentleman of Philadelphia County" wrote in his *American Shooter's Manual* that the country was one "destitute of game laws, and almost without any legal restrictions" on the destruction of game.[96]

The increasing scarcity of important species was well observed before 1850, and early legislation at least paid lip service to the notion that abundance was preferred to scarcity. But although the passage of laws was relatively costless, their enforcement required real resources, threatened customary free access to wildlife, and questioned the sanctity of private property in land. These "costs" notwithstanding, the next half-century saw the development of a powerful sentiment favoring protection of wildlife and witnessed the growth of state regulation designed to achieve that goal.

NOTES

1. Anonymous, *A Description of Georgia, by a Gentleman Who Has Resided There Upwards of Seven Years, and Was One of the First Settlers,* 1741. Reprinted in Peter Force, *Tracts and Other Papers Relating Principally to the Origin, Settlement, and Progress of the Colonies in North America, from the Discovery of the Country to the Year 1776,* 4 vols. (Washington, D.C.: Peter Force, 1838), II, 3-4.

2. Phillip Alexander Bruce, *Economic History of Virginia in the Seventeenth Century,* 2 vols. (New York: Macmillan & Co., 1896; reprinted by the Johnson Reprint Corp., New York, 1966), I, 124.

3. John Josselyn, *New England's Rarities* (1673), cited in James B. Trefethen, *An American Crusade for Wildlife* (New York: Winchester Press, 1975), p. 38. Not contained in the edition of Josselyn cited in note 57 below.

4. For example, John Brickell's *The Natural History of North Carolina* was written to advertise the lands of Earl Granville [William Martin Smallwood, *Natural History and the American Mind* (New York: Columbia University Press, 1941), pp. 23-24]. See also Hans Huth, *Nature and the American: Three Centuries of Changing Attitudes* (Berkeley: University of California Press, 1957), p. 4.

5. In 1788, the legislature of the short-lived region of Franklin, now a part of Tennessee, established that the governor of the territory would receive 1,000 deer skins annually; the chief justice and the

attorney general, 500 each; the secretary to the governor would receive 500 raccoon skins; the treasurer 450 otter; the county clerks, 300 beaver; the clerk of the House of Commons, 200 raccoon; members of the assembly, 3 raccoons per day; the fee to a justice for signing a warrant, 1 muskrat; and to a constable for serving a warrant, 1 mink [Henry Chase, *Game Protection and Propagation in America* (Philadelphia: J. B. Lippincott Company, 1913), p. 40].

6. Thomas Nuttall, *A Manual of Ornithology: The Land Birds* (Boston: Hilliard, Gray, and Co., 1840), p. 800.

7. Wilson's curious calculation derived from his interest in the total food consumption of the huge flocks and their impact on domestic crops. At one-half pint per bird per day, the observed flock would consume 17,420,000 bushels per day, an amount which, at an annual rate, roughly equals the combined total U. S. output of corn and wheat in 1970. Brewer's edition of Wilson includes a passage from Audubon's later work that reports an 1813 flock of pigeons that was estimated to contain exactly one-half the number of birds observed by Wilson, consuming, coincidentally, exactly one-half the amount of grain [Alexander Wilson, *American Ornithology*, ed. T. M. Brewer (Boston: Otis, Broaders, and Company, 1840), pp. 393-94, 399. See also Chapter 3; note 88.

8. Letter to William Hornaday, quoted in Frank Gilbert Roe, *The North American Buffalo: A Critical Study of the Species in its Wild State*, 2d ed. (Toronto: University of Toronto Press, 1970), pp. 358-59. Hornaday, noting that the bison travels in a wedge-shaped herd, reduced Dodge's estimate by two-thirds.

9. James B. Trefethen, "Wildlife Regulation and Restoration," in Henry Clepper, ed., *Origins of American Conservation* (New York: The Ronald Press Co., 1966), p. 18.

10. See the discussion on the relationship between land-use changes and wildlife abundance in Chapter 3.

11. Edwin Bottomley, *An English Settler in Pioneer Wisconsin: The Letters of Edwin Bottomley, 1842-1850*, ed. and with an Introduction by Milo M. Quaife, in State Historical Society of Wisconsin *Collections* (Madison, 1918) XXV, 90-91. See also Arthur Alphonse Ekirch, Jr., *The Idea of Progress in America, 1815-1860* (New York: Columbia University Press, 1944), pp. 91-92. For a more cynical view on wildlife abundance, see Jared Van Wagenen, Jr., *The Golden Age of Homespun* (New York: Hill and Wang, 1953), p. 92.

12. Although native American populations appeared to have occupied much less disruptive niches than did those who followed, they cannot claim immunity of the charge of wildlife exploitation. On

man's role in the Pleistocene extinctions, see Paul S. Martin, "Prehistoric Overkill," Arthur J. Jellinek, "Man's Role in the Extinction of Pleistocene Faunas," and James J. Hester, "The Agency of Man in Animal Extinctions," in P. S. Martin and H. E. Wright, Jr., eds., *Pleistocene Extinctions: The Search for a Cause* (New Haven: Yale University Press, 1967). On more recent interactions, see James Tober, "The Allocation of Wildlife Resources in the United States, 1850-1900" (Ph.D. diss., Yale University, 1973), pp. 9-12.

13. Roderick Nash, *Wilderness and the American Mind* (New Haven: Yale University Press, 1967), p. 26. On Old World roots and New World wilderness conditions, see Nash, pp. 1-43, passim.

14. The origins and vitality of this dream, at least through the mid-nineteenth century, are the themes of Leo Marx, *The Machine in the Garden: Technology and the Pastoral Ideal in America* (New York: Oxford University Press, 1964).

15. Trefethen, *American Crusade*, pp. 55-58.

16. Winthrop's introspection revealed additional costs. Fowling, for him, was too time-consuming, strenuous, dangerous, and expensive. Edmund S. Morgan, *The Puritan Dilemma: The Story of John Winthrop* (Boston: Little, Brown and Company, 1958), pp. 9-10. John C. Miller, *The First Frontier: Life in Colonial America* (New York: Delacorte Press, 1966), p. 55.

17. Herbert Manchester, *Four Centuries of Sport in America, 1490-1890* (New York: The Derrydale Press, 1931), pp. 17-18.

18. The Reverend Joseph Seccomb, "Business and Diversion inoffensive to God, and necessary for the Comfort and Support of human Society," sermon delivered at Ammauskeeg Falls, New Hampshire, 1739 (1743; reprint ed. Manchester, N.H.: n.p., 1892), pp. 12-13.

19. Several recent town histories describe these dynamics. See, for example, Kenneth A. Lockridge, *A New England Town: The First Hundred Years, Dedham, Massachusetts, 1636-1736* (New York: W. W. Norton, 1970).

20. The religious man would not engage in Sunday fishing, but it was "with a touch of envy as well as stern disapproval that the Bible-reading skipper watched on a Sabbath morning his rivals haul in cod after cod while his own lines were idle." [Thomas Jefferson Wertenbaker, *The Puritan Oligarchy: The Founding of American Civilization* (New York: Charles Scribner's Sons, 1947), p. 195].

21. Joseph W. Moulton, *An Address Delivered at St. Paul's Church, Buffalo, on the Anniversary Celebration of the Niagara and Erie Society for Promoting Agriculture and Domestic Manufactures* (Buffalo: D. M. Day and H. A. Salisbury, 1821), p. 5.

22. The 1817 remarks of President Monroe, quoted in Alvin M. Josephy, *The Indian Heritage of America* (New York: Alfred A. Knopf, 1968), p. 334.

23. J. Hector St. John Crèvecoeur, *Letters from an American Farmer* (New York: Fox, Duffield and Co., 1904), p. 59.

24. Ray Allen Billington, *The Far Western Frontier, 1830-1860* (New York: Harper and Row, 1956), pp. 44, 49, 51. Emphasis added.

25. Massachusetts, Commissioners on the Zoological and Botanical Survey, *A Report on the Quadrupeds of Massachusetts Published Agreeably to an Order of the Legislature by the Commissioners on the Zoological and Botanical Survey of the State*, by Ebenezer Emmons (Cambridge, Mass.: Folsom, Wells, and Thurston, 1840), p. 77. Emmons submitted a similar report in 1838 in which he limits his remarks largely to domesticated animals. In his brief remarks on "a few of the rarer animals" of the state, he notes that the moose, because it has been "domesticated and broken to the harness," may be kept from extinction if "enterprising persons should anticipate a profit from . . . raising them for the value of their meat." An accompanying report on birds of the state by the Reverend W. O. B. Peabody concludes that birds have diminished as they have become less necessary and that "legislation will not be able, even were it worthwhile, to preserve them" [*Reports of the Commissioners on the Zoological Survey of the State* (Boston: Dutton and Wentworth, State Printers, 1838), pp. 29, 31].

26. Thomas Jefferson, *Notes on the State of Virginia*, edited with an Introduction and Notes by William Peden (Chapel Hill: University of North Carolina Press, 1955), Query XIX, p. 165.

27. Letter to Benjamin Austin, cited in Marx, *Machine in the Garden*, p. 139.

28. Ralph Waldo Emerson, *The Complete Works of Ralph Waldo Emerson*, with a Biographical Introduction and Notes by Edward Waldo Emerson, 12 vols. (Boston and New York: Houghton, Mifflin and Company, 1903), I, 391. See also Ekirch, *The Idea of Progress in America*, p. 160.

29. Cited in Ekirch, *The Idea of Progress in America*, p. 129.

30. Ernest S. Englebert, "American Policy for Natural Resources: A Historical Survey to 1862" (Ph.D. diss., Harvard Univesity, 1950), pp. 258-60. Twentieth-century views on resource abundance are surveyed in Harold J. Barnett and Chandler Morse, *Scarcity and Growth: The Economics of Natural Resource Availability* (Baltimore: Johns Hopkins University Press, for Resources for the Future, 1963), and V. Kerry Smith, ed., *Scarcity and Growth Reconsidered* (Baltimore: Johns Hopkins University Press, for Resources for the Future, 1979).

31. Englebert, "American Policy for Natural Resources," p. 133.

32. Nash, *Wilderness and the American Mind*, p. 67. On the Romantic wilderness and its American discovery, see Nash, pp. 44-83, passim.

33. Joseph Kastner, *A Species of Eternity* (New York: E. P. Dutton, 1978), p. 123.

34. Ibid., p. 124. A great deal of questionable natural history accumulated during this period, not all of which is readily attributable to such political motives. An example is the wonderfully fantastic story of the beaver, preserved by Beltrami, who reports only "what I have myself actually seen, and been from good authority informed of...." The species is

> divided into tribes, and sometimes merely into small bands, each of which has its chief; and order and discipline exist in these distinct societies to a greater extent probably than among the Indians, or even among some civilized and polished nations.... Each tribe has its peculiar territory. If any foreigner be taken in the act of marauding, he is delivered over to the chief, who, on the first offense, chastises him with a view to correction; but, for the second, deprives him of his tail, which is considered as the greatest disgrace to which a beaver can be exposed; for the tail is the carriage on which he conveys stones, mortar, provisions, etc., and it is also the trowel (the figure of which it represents exactly) which he uses in building. [G. C. Beltrami, *A Pilgrimage in Europe and America, Leading to the Discovery of the Sources of the Mississippi and Bloody River*, 2 vols. (London: Hunt and Clarke, 1828), II, 426-27].

On the "Fabulous History of the Beaver," see John D. Godman, *American Natural History*, 3d. ed., 2 vols. (Philadelphia: Hogan and Thompson, 1836), I, 280-91.

35. Nash, *Wilderness and the American Mind*, p. 68. Kastner, *Species of Eternity*, pp. 124-27.

36. On Charles Willson Peale's Philadelphia Museum, see Kastner, *Species of Eternity*, pp. 143-58. On the career of P. T. Barnum and the fine line between science and showmanship, see Neil Harris, *Humbug: The Art of P. T. Barnum* (Boston: Little, Brown and Company, 1973). Of special interest is the Smithsonian Institution, established in 1846, in its role as sponsor of the federal government's scientific inquiries and a repository for its collections. See Paul H. Oehser, *The Smithsonian Institution* (New York: Praeger Publishers, 1970), and Geoffrey T. Hellman, *The Smithsonian: Octopus on the Mall* (Philadelphia: J. B. Lippincott Company, 1967).

37. John Chancellor, *Audubon: A Biography* (New York: The Viking Press, 1978), p. 196. "And all this because the World is all agog—for what? for *Bugs* the size of *Water Melons*... I almost wish I could be turned into a *Beetle* myself" (ibid., p. 199).

38. On the development of American natural history, see Smallwood, *Natural History and the American Mind* (New York: Columbia University Press, 1941); Kastner, *A Species of Eternity*; Wayne Hanley, *Natural History in America, from Mark Catesby to Rachel Carson* (New York: Quadrangle/New York Times Book Company, 1977); and Donald Culross Peattie, *Green Laurels: The Lives and Achievements of the Great Naturalists* (New York: Simon and Schuster, 1936). On natural history illustrators, see S. Peter Dance, *The Art of Natural History: Animal Illustrators and Their Work* (Woodstock, N.Y.: Overlook Press, 1978).

39. Nash, *Wilderness and the American Mind*, pp. 101-2 and pp. 96-121, passim. See Chapter 4 below for a discussion of the preservation of wildlife on the land.

40. Thomas Morton, *New English Canaan; or New Canaan, Containing an Abstract of New England* (1632). *Tracts and Other Papers*, collected by Peter Force, 4 vols. (Washington, D.C.:, Printed by Peter Force, 1838), II, 52.

41. Stanley P. Young, "The Deer, the Indians, and the American Pioneers," in Walter P. Taylor, ed., *The Deer of North America* (Harrisburg, Pa., and Washington, D.C.: The Stackpole Company and the Wildlife Management Institute, 1956), p. 21.

42. Connecticut, *The Public Records of the Colony of Connecticut*, ed. J. Hammond Trumbull and Charles J. Hoadley, 15 vols. (Hartford: 1850-90), II, 308; III, 31, 78.

43. Verner W. Crane, *The Southern Frontier, 1670-1732* (Philadelphia: University of Pennsylvania Press, 1929). The average export to England between 1699 and 1715 was 54,000 skins, with a peak of 121,355 in 1707. The volume of trade fluctuated throughout the remainder of the colonial period according to the character of relations with the Indians and reached a high of 160,000 skins in 1748. Crane regards the significance of this trade not in the substantial revenues that were brought into the colony but rather in the good relations that were maintained with neighboring Indians (pp. 108-112).

44. John D. Godman, *American Natural History*, II, 120.

45. The literature of the fur trade is voluminous. A standard work is Hiram Martin Chittenden, *The American Fur Trade of the Far West*, 2 vols. (New York: The Press of the Pioneers, 1935). The renewed interest in the Native American role in the trade is represented by

Arthur J. Ray, *Indians in the Fur Trade: Their Role as Trappers, Hunters, and Middlemen in the Lands Southwest of Hudson Bay, 1660-1870* (Toronto: University of Toronto Press, 1974), and Calvin Martin, *Keepers of the Game: Indian-Animal Relationships and the Fur Trade* (Berkeley: University of California Press, 1978).

46. See Curtis P. Nettels.*The Emergence of a National Economy* (New York: Harper and Row, 1962), pp. 209-16, and Billington, *The Far Western Frontier*, pp. 41-68.

47. William Cobbett, *A Year's Residence in the United States of America*, 2d ed., (London: Sherwood, Neely & Jones, 1819), pp. 371-72.

48. *The Sportsman's Companion: Or, An Essay on Shooting* (New York: Robertsons, Mills, and Hicks, 1783), pp. 29-30. According to Manchester, "A Gentlemen" was a member of the British Army in New York (*Four centuries of Sport in America*, pp. 17-18).

49. Richard C. Wade, *The Urban Frontier* (Chicago: University of Chicago Press, 1959), pp. 312-13.

50. Foster Rhea Dulles, *America Learns to Play: A History of Popular Recreation, 1607-1940* (1940; reprint ed. Gloucester: Peter Smith, 1959), p. 71.

51. Ibid., p. 72.

52. George Catlin, hunting grouse with the Infantry on the Lower Mississippi, "followed the sportman's style for a part of the afternoon," but, as the birds were driven in great numbers before a prairie fire, he was "quite ashamed to confess" that "we murdered the poor birds" [George Catlin, *Letters and Notes on the Manners, Customs, and Conditions of the North American Indians*, 2 vols. (1841; reprint ed. (Minneapolis: Ross and Haynes, 1965), II, 16].

53. Americans customarily chose utility over style, and, as Rosenberg observes, the corollary willingness to sacrifice control over production to the needs of a nascent industrial system was an important element in nineteenth century growth. Rosenberg finds a pertinent example in firearms where the "British civilian market was long dominated by peculiarities of taste which essentially precluded machine techniques," whereas American consumers appeared willing to accommodate their tastes to the machine [Nathan Rosenberg, *Technology and American Economic Growth* (New York: Harper & Row, 1972), pp. 43-51].

54. Huth, *Nature and the American*, p. 55.

55. From an 1841 letter of Dr.Arnold to Mr. Justice Coleridge, quoted in Charles Eliot Goodspeed, *Angling in America: Its Early History and Literature* (Boston: Houghton Mifflin Company, 1939), p.

64, note 2. Sport as recreation gained favor by 1850 as cities grew large and crowded. William H. Schreiner, in *Schreiner's Sporting Manual* (Philadelphia: S. D. Wyeth, 1841), cites the "well established fact that by a judicious indulgence in the favorite sciences of fishing and gunning, many important diseases may be removed; among which are cited indigestion, nervous derangement, nervous pains, and debility, rheumatism, and spinal affections, and even the indomitable, through very prevalent and alarming complaint of the lungs" (quoted in Goodspeed, *Angling in America*, p. 158).

56. Species extinct by 1850 or to become extinct shortly thereafter include the Labrador duck (*Camtorhynchus labradorium*), the great auk (*Alca impennis*), the Carolina parakeet (*Conuropsis carolinensis*), and the sea mink (*Mustela macrodon*). Terrestrial mammalian extinctions were limited to races of species extant in other region. They include the Eastern bison (*Bison bison pennsylvanicus*) and the Eastern elk (*Cervus canandensis canandensis*). The status of the eastern cougar (*Felis concolor couguar*) long thought to be extinct, is presently under investigation. See Michael Frome, "Panthers Wanted—Alive, Back East Where They Belong" *Smithsonian* 10 (June 1979), 83-87. A large number of other species, including fur bearers, large mammals, and game birds, were locally extirpated. For a detailed inventory of species, see James C. Greenway, *Extinct and Vanishing Birds of the World*, Special Publication No. 13 (New York: American Committee for International Wildlife Protection, 1958); Vinzenz Ziswiler, *Extinct and Vanishing Animals*, rev. English ed. by Fred and Pille Bunnell (New York: Springer-Verlag, 1967); Glover M. Allen, *Extinct and Vanishing Mammals of the Western Hemisphere*, Special Publication No. 11 (New York: American Committee for International Wildlife Protection, 1942); Peter Mathiessen, *Wildlife in America* (New York: The Viking Press, 1959); and Paul Ehrenfeld, *Biological Conservation* (New York: Holt, Rinehart and Winston, 1970).

57. *New England's Rarities Discovered* (1672). *Archaeologia Americana. Transactions and Collections of the American Antiquarian Society*, 12 vols. (Worcester: Printed for the Society, 1820-1911), IV, 144.

58. Quoted in E. Douglas Branch, *The Hunting of the Buffalo* (New York: D. Appleton and Co., 1929), p. 95.

59. J. A. Allen, "The Decrease of Birds in Massachusetts," *The Bulletin of the Nuttall Ornithological Club* 1 (September 1876), 54.

60. Quoted in William T. Hornaday, *Our Vanishing Wildlife: Its Extermination and Preservation* (New York: New York Zoological Society, 1913), frontispiece.

61. David Potter, *People of Plenty: Economic Abundance and the American Character* (Chicago: University of Chicago Press, 1954), pp. 91, 119.

62. Englebert, "American Policy for Natural Resources," pp. 190-93. See also Joseph Ellison, "The Mineral Land Question in California, 1848-1866," *Southwestern Historical Quarterly* 30 (1926), 34-55, reprinted in Vernon Carstensen, ed., *The Public Lands: Studies in the History of the Public Domain* (Madison: University of Wisconsin Press, 1968), pp. 71-92; and Gary D. Libecap, "Economic Variables and the Development of the Law: The Case of Western Mineral Rights," *Journal of Economic History* 38 (June 1978), 338-62.

63. Englebert, "American Policy for Natural Resources," pp. 342-46, and Lucile Kane, "Federal Protection of Public Timber in the Upper Great Lakes States," *Agricultural History* 23 (1949), 135-39, reprinted in Carstensen, *The Public Lands*, pp. 439-47. For a broad overview of natural resource exploration, appropriation, exploitation, and conservation, see William N. Parker, "The Land, Minerals, Water, and Forests," in *American Economic Growth, An Economists' History of the United States* (New York: Harper and Row, 1972), pp. 93-120.

64. The similarities were recognized in *Westmorland and Cambria Natural Gas Co. v. De Witt et al.*, 130 Pa St. 235 (1889). "Water and oil, and still more strongly gas, may be classed by themselves, if the analogy be not fanciful, as minerals *ferae naturae*. In common with animals, and unlike other minerals, they have the power and the tendency to escape without the volition of the owner."

65. William F. Schulz, Jr., *Conservation Law and Administration: A Case Study of Law and Resource Use in Pennsylvania* (New York: Ronald Press, 1953) pp. 45-46.

66. Fearing that not all states might give such guarantees or anticipating a time when a federal legislature might attempt to limit the rights of states in allocating access to wild animals, the minority of the Constitutional Convention of Pennsylvania recommended that the United States Constitution guarantee these rights to hunt and fish unrestrained "by any laws to be passed by the Legislature of the United States" [*Address and Reasons of Dissent of the Minority of the Convention of the State of Pennsylvania to their Constituents*, December 12, 1787 (Evans Microprint No. 20618), and Thomas A. Lund, "Early American Wildlife Law," *New York University Law Review* 51 (November 1976), 712, note 76].

67. 68 Vt. 353-54, 358 (1895).

68. Lawrence M. Friedman, *A History of American Law* (New York: Simon and Schuster, 1973), prologue and part I, passim. See also Grant Gilmore, *Ages of American Law* (New Haven: Yale University Press, 1977), pp. 19-25.

69. William Nelson's investigation into 2,784 criminal prosecutions in seven Massachusetts counties between 1760 and 1774 reveals that 13 percent were offenses against property (for example, robbery, burglary, larceny), 15 percent were crimes of violence (for example, assault, homicide, trespass, disturbing the peace), 38 percent were sex crimes (for example, adultery, cohabitation, prostitution), 13 percent were religious offenses (for example, blasphemy, missing church, profanity), and 20 perent were miscellaneous offenses including the unseasonable killing of deer, violation of liquor licensing laws, fraud in the marketplace, drunkennes, and libel [William E. Nelson, *Americanization of the Common Law: The Impact of Legal Change on Massachusetts Society, 1760-1830* (Cambridge, Mass.: Harvard University Press, 1975), pp. 36-45]. Enforcement of the pre-Revolutionary criminal law served not to "segregate and punish a distinct, identifiable, and downtrodden criminal class," but rather to reintegrate those who had violated "puritanical religious and moral standards . . . into the existing community structure" (ibid., p.40).

70. Nelson, *Americanization of the Common Law*, p. 29, and Chapter 2, passim.

71. See Chapter 4, text at notes 46-47 and Chapter 6, text at notes 156-59.

72. Friedman, *History of American Law*, pp. 205-8. See also Richard L. Bushman, *From Puritan to Yankee* (Cambridge, Mass.: Harvard University Press, 1967), pp. 41-103.

73. Morton J. Horwitz,"The Transformation in the Conception of Property in American Law, 1780-1860," *University of Chicago Law Review* 40 (1973), reprinted in Lawrence M. Friedman and Harry N. Scheiber, eds., *American Law and the Constitutional Order: Historical Perspectives* (Cambridge, Mass.: Harvard University Press, 1978), p. 144.

74. Morton J. Horwitz, *The Transformation of American Law 1780-1860* (Cambridge, Mass.: Harvard University Press, 1977), p. 66, and Chapter 3, passim.

75. Ibid., p. 99.

76. Gilmore, *Ages of American Law*, Introduction.

77. 65 N.C. 315 (1871). See also the 1850 Massachusetts statute (*Statutes*, p. 474) providing that "the taking, without consent of the

owner and with a felonious intent, of any beast or bird ordinarily kept in a state of confinement, and not being the subject of larceny at common law, shall be held to be larceny. . . . "

78. Gilmore, *Ages of American Law*, pp. 10-11, 25-36. "Even though the idea of a wholesale federalization of the substantive law was rejected at an early stage, the Supreme Court has, throughout our history, discovered and exploited various methods of establishing federal supremacy—and thus national uniformity" (ibid., pp. 29-30).

79. See Chapter 5 below on the legal history of wildlife and Chapter 6 on nineteenth-century federal wildlife law.

80. See Lund, "Early American Wildlife Law," for "an eclectic collection of early American wildlife laws."

81. Connecticut, *The Public Records of the Colony of Connecticut*, I, 149, 367, 377; II, 61; IV, 287; V,406; VII, 209; VIII, 572; IX, 277; X,78. The bounty, repealed in 1774 (Ibid., XIV, 330), was reintroduced by the State of Connecticut in 1784 [Connecticut, *Records of the State of Connecticut*, 11 vols. edited by Charles J. Hoadly, Leonard Woods, and Albert E. Van Dusen. (Hartford, 1894-1967), I, 149, 448-9; II, 70, 337; V, 342-7.

82. Stanley Paul Young, *The Wolf in North American History* (Caldwell, Idaho; The Caxton Printers, Ltd., 1946), pp. 74-75, and Allen, *Extinct and Vanishing Mammals*, p. 211. The principle of equal responsibility for noxious animals was again illustrated in Ohio when, during an eruption of the squirrel population in 1808, the legislature required each free male to supply one hundred squirrel scalps to the county clerk or pay three dollars. [D. B. Warden, *A Statistical, Political, and Historical Account of the United States of America*, 3 vols. (Edinburgh; Archibald Constable & Co., 1819), II, 243-44]. Similarly, the states of Missouri and Nebraska passed legislation in 1877 requiring ablebodied men to fight the war against the grasshopper. The Missouri law (*Statutes*, p. 336-37) required two days' service upon a penalty of one dollar per day, and the Nebraska law (*Statutes*, pp. 154-55) required up to twelve days' service per year, upon a penalty of ten dollars.

83. Connecticut, *The Public Records of the Colony of Connecticut*, I, 283; V, 563-4.

84. Ibid., IV, 248-9,345; V, 524-5; VI, 28-9; VIII, 268.

85. Massachusetts, *Acts and Resolves, Public and Private, of the Province of the Massachusetts Bay*, 21 vols. (Boston: Wright and Potter, Printers to the State, 1869-1922), I, 153; II, 988-90. New Hampshire, *Laws of New Hampshire*, ed. Albert Stillman Batchellor and H. H. Metcalf, 7 vols. (Manchester, N.H.: John B. Clarke Co., 1904-1918), II,

585-6. On the extensive exchange of legal code among jurisdictions, see Friedman, *History of American Law*, pp. 79-81.

86. New York, *The Colonial Laws of New York from the Year 1664 to the Revolution*, 5 vols. (Albany: J. B. Lyon, State Printer, 1894), III, 618-20.

87. U.S. Department of Agriculture, Bureau of Biological Survey, *Chronology and Index of the More Important Events in American Game Protection, 1776-1911*, by T. S. Palmer, Bulletin No.41 (Washington, D.C.: G.P.O., 1912).

88. See Lund, "Early American Wildlife Law," pp. 716-19, 721.

89. Connecticut, *The Public Records of the Colony of Connecticut*, XII, 498-9; XIII, 118, 401-2.

90. Lund, "Early American Wildlife Law," p. 721, note 144. See Chapter 5, note 45.

91. Nelson, *Americanization of the Common Law*, p. 123. This is the controversy addressed in *Trout and Salmon Club* v. *Mather*. See text at note 67.

92. Helenette Silver, *A History of New Hampshire Game and Furbearers* (New Hampshire: Fish and Game Department, Survey Report No. 6, 1957), p. 195.

93. George Bird Grinnell, "American Game Protection: A Sketch," in Grinnell and Charles Sheldon, ed., *Hunting and Conservation* (New Haven: Yale University Press, 1925), p. 221.

94. Henry Chase, *Game Protection and Propagation in America* (Philadelphia: J. B. Lippincott Company, 1913), p. 38. The importance of equal access to game was spelled out by John Lawson in his eighteenth-century work on the Carolinas. In that region,

> property hath a large scope, there being no strict laws to bind our privileges. A qust [sic] after game being as freely and peremptorily enjoyed by the meanest planter, as he that is the highest in dignity, or wealthiest in the province. Deer and other game that are naturaly wild, being not immured, or preserved within boundaries, to satisfy the appetite of the rich alone. A poor laborer that is master of his gun, etc., hath as good a claim to have continued coarses [sic] of delicacies crowded upon his table, as he that is master of a great purse. [John Lawson, *The History of Carolina* (Raleigh: Strother and Marcom, 1860), p. 29].

95. Cobbett, *A Year's Residence*, p. 376. Against his travels in America, Audubon was distinctly impressed by the number of "trespassers will be prosecuted" signs which he encountered on an 1826 tour of England. The country was "all hospitality *within* and ferocity *without*.

No one dare *trespass*, as it is called. Signs of *large dogs* are put up; steel traps and spring guns are set up, and even *eyes* are kept out by high walls" (Chancellor, *Audubon*, p. 122).

96. A Gentleman of Philadelphia County, *American Shooter's Manual* (New York: Ernest R. Gee, 1928; originally published, 1827), p. vii.

2

SPORTSMEN AND OTHER HUNTERS

> Every school-boy knows that association for every purpose
> of human action is a characteristic of the age in which we
> live, and that while each man in his sphere may do much by
> precept and example, many men associated can accomplish
> a vast deal more than they can do separately.
>
> *American Sportsman*, 1875

The casual reception accorded the decline of wildlife during
the early phases of American expansion and growth was transformed, during the second half of the nineteenth century, into
an active battle over access to remaining populations. Three
groups—sportsmen, market hunters and game dealers, and
landowners—emerged to play strategic roles. Each sought to
preserve traditional patterns of interaction with wildlife and to
remain free to develop new patterns as they seemed advantageous. But the diversity of demands for wildlife, combined
with increasingly limited populations, precluded mutual satisfaction. The sport hunter's favorite retreat and the market
hunter's lucrative haunt were more likely to coincide, and
either was more likely now to be enclosed by a farmer's fence
or buried in the rubble of a timber harvest.

Were property rights in wildlife clearly articulated and widely
supported, the conflict need not have persisted. Some claims
to wildlife would have taken precedence over others, and
allocation to the limited resource would have been regulated
through voluntary exchange and bureaucratic control. But it
was this very conflict that created the need for increased speci-
fication of property rights, a conflict that emerged when the

traditional patterns of resource use were threatened by scarcity, changing demands, and new technologies.

The task in this chapter is to characterize each of these major groups and to outline their roles in this process.* The most stylized version finds sportsmen, practicing the healthful and refined art of taking game in moderation; members of the game industry, engaging in free enterprise to meet the legitimate needs of a growing and hungry nation; and landowners, husbanding the land as private property in the name of national prosperity. Each group laid claim to a broad slice of American history and culture, and each drew broad support. Each molded public opinion and was molded by it.

This stark portrayal does not imply that the division was clear or widely accepted either by the parties to the dispute or by the general public. It plays in favor of sportsmen whose strength lay in part in their ability to create a self-righteous identity and against other hunters who often sought to obscure that identity. The division also throws together a disparate band of landowners and users, including farmers, cattlemen, miners, and loggers, most of whom doubtlessly hunted for subsistence or for the market but whose relationship to the land did not owe its origins to wildlife.

This tripartite division also leaves out a role for other constituencies—fashion-conscious women who first clamored for and later scorned milliners' ornate arrangements of birds and feathers, urban epicures who sought out woodcock, canvas back, and venison over and under the counter, and naturalists who collected extensively in the field and acquired a broad knowledge of the American environment and its resources. Throughout the period of this study, a more pervasive public sentiment ebbed and flowed and was instrumental in supporting the positions of primary groups and in creating a climate of opinion that these groups could selectively exploit for their own purposes. The sentiment peaked over issues such as the hunting of the bison, the use of passenger pigeons in trap shooting, and the employment of nongame bird plumage in

---

*Another characterization, one that depends on analytical rather than on historical argument, may be found in Appendix 1.

the millinery trade. These were instances in which the questioned behavior was highly visible, associated with the provision of what were viewed as nonessential goods, evocative of humanitarian feelings, and for which the primary actors were thought to be easily identifiable. The role of these wider interests will be considered as appropriate in later chapters.

THE SPORTSMAN

Sport hunting did not attract a large following prior to 1850. But the increase in leisure accompanying increased incomes and the urbanization which removed from everyday experience a contact with the natural environment combined to change that: urbanization made attractive a renewal of the bonds with nature, and leisure time made it possible. The tradition of the English sporting gentleman provided the vehicle.

More than any other individual, Henry William Herbert—or Frank Forester, as he was popularly known—was responsible for making available to the class of potential sportsmen in the United States the accumulated wisdom and protocol of the British nimrod.[1] Beginning with *Field Sports in the United States, and the British Provinces of America*, published in two volumes in 1848, and continuing with a number of other works, Forester set the mode for the man who desired to become a "sportsman."[2] There was little doubt in his mind that field sports had enabled the British nobility to retain a physical stature and strength greater than that of its peasantry and to resist the "deteriorating influences of wealth, luxury, and breeding-in-and-in."[3] The American ruling class was at a decided disadvantage in its distance from the invigorating rural setting, and its members were well advised to take up the sporting life without delay.

Since the hunt may be undertaken by any individual with a modicum of skill and the rudest of implements, the true sportsman, to retain his identity, must proceed in a highly stylized manner. He acquired a specialized vocabulary, a deep interest in natural history, a code of ethics, and fashionable dress. The definitive vocabulary for the American sportsman is found in Forester's *Field Sports*, in which one learns the proper collective terms for wildlife: a "whiteness" of swans, a "gaggle" of geese, a "gang" of brant, a "team" of ducks (for a large number) and a "plump" of ducks (for a small number), a "company" (or "trip") of widgeon, a "flock" of teal, a "whisp" of snipe, and a "sege" of bitterns or heron. One learns, too, that while he may refer to a "herd" of bison, moose, or caribou, he must note a "gang" of elk and a "drove" of wolves. Finally, two grouse, quail, or hares are a "brace," and three are a "leash," but whereas two woodcocks, plover, or waterfowl are a "couple," three are a "couple-and-a-half," and "the applying of these terms *vice versa* is a bad sporting blunder."[4]

Elisha J. Lewis berated "many of our sporting acquaintances" for being "woefully deficient" in a knowledge of this vocabulary, which led to their making "the most egregious blunders in their vain efforts to appear 'au fait' in all that pertains to the dog and the gun."[5] In his *Hints to Sportsmen*, the first major work built on Forester's foundation, and which was to become increasingly popular in later editions as *The American Sportsman*, Lewis elaborated the list of correct terminology available to the general readership. In commanding a dog, for example, "toho!" is the term employed to make pointers or setters "come to a stand" and "Seek dead! find dead bird!" means just what it says. Among the sportsman's "egregious blunders" might have been to refer to a "pair of setters" in the field, being unaware that "pair" refers to two individuals of the same species joined by nature, as a male and a female, whereas a "couple" or "brace" is an involuntary union, and consequently would have been proper.[6]

The sportsman's manner in the field was likewise well defined. Deer were to be stalked and not shot from platforms placed above known runs, nor driven onto crusted snow, nor into water where they are no match for a hunter in a boat. Upland game birds were to be shot on the wing. The sportsman should remember that "all birds that cross belong exclusively to that person to which side they bear" and that "shots should be taken alternately, when as fair for one as for the other."[7] The sportsman who happened to fire with his partner was instructed, if the bird fell, never to intimate that he killed it. "If your dog retrieves, offer it to your guest, and if you killed

it, unless he is a hog, he will say so."[8] Seneca (the pen name of
H. H. Soulé) offered other rules of behavior: "Always be polite
and unselfish. . . . Drink little or no liquor. . . . If you take down
bars, put them up again. . . . Walk abreast of your associate,
never ahead or behind. . . . Don't lie about your exploits."[9]

Lewis was critical of the "snap shot," who pulled the trigger
too often, and of the "poking shot" who pulled it not often
enough. But forced to make a choice, "we would rather miss
three shots out of five all day long than go pottering about."[10]
Similarly, a correspondent to the American Sportsman offered
more credit to the hunter "who walks three miles and kills
eighteen birds out of twenty" than to the hunter "who tramps
all day and blunders down fifty wing-tips, missing at every
other shot."[11] Apparently, it was style rather than shooting
frequency that marked the sportsman. One who goes "pottering about," who "tramps," or who "blunders down" is surely
no sportsman in the sense in which these hunters sought to
distinguish themselves.

The ideal sportsman was a sportsman/naturalist. Forester
blamed half of the difficulty that arose among sportsmen on
the "stupid misnomers" that had become attached to the
species in the United States, and, to improve communications,
he promoted the use of Linnaean binomials.[12] Lewis, who
claimed that "no one can expect to become an accomplished
sportsman without studying very closely the individual characteristics of every species of game that he pursues,"[13] prefaced Hints to Sportsmen with a glossary of ornithological terms.
Indeed, many of the most important nineteenth-century contributions to the natural history of game species were made by
those who would have considered themselves sportsmen first
and naturalists only by experience. An example was John Dean
Caton who, in his highly regarded Antelope and Deer of America,
observed that the sportsman, when he kills an animal, "takes it
up and examines it as he would a book full of knowledge."[14]

The characteristics of the true sportsman were offered, often
in contrast to those of other hunters, throughout the literature.
Robert Roosevelt distinguished him from the "daring back-
woods man of the Far West," who is "less a sportsman than a
mighty hunter," from the market gunner who "shoots with a

view of selling his game," and from the pot-hunter "who kills that he may eat." The sportsman, on the other hand, "pursues his game for pleasure; he ... makes no profit of his success ... shoots invariably upon the wing, and never takes mean advantage of bird or man."[15] Others were wont to use somewhat more emotional language, both in describing the sportsman and in denouncing the sham. Bumstead had no time for pot hunters who snare birds "for the sake of filthy lucre,"[16] and Lewis described the pot hunter as "the most disgusting, the most selfish, the most unmanly, the most heartless ... of all the disagreeable characters that a well-bred sportsman is likely to be thrown into contact with."[17] The sportsman, on the other hand, embodied all good qualities. His sociability was among his grandest attributes—"How it shines forth in the wilderness, brightening up the camp with its beaming effulgence."[18]

The sportsman's ideology and public image were simultaneously created and reinforced by the rapidly growing sporting press, led by several influential and widely distributed periodicals. The foremost of those was Charles Hallock's *Forest and Stream*, which commenced weekly publication in New York on August 14, 1873. Ownership passed in 1879 to George Bird Grinnell who outspokenly led the journal through the remainder of its thirty-eight-year run.[19] G. W. Martin, in his presidential address to the membership of the Kennebec Association for the Protection of Game, attributed American interest in sport to *Forest and Stream*, in conjunction with Frank Forester and the Smithsonian Institution. "Mr. Hallock would feel justly proud, did he know how much valuable information the true sportsmen of the Kennebec acknowledge to have received from him."[20]

But in spite of the best efforts of the sporting press to preserve a special place for these most highly principled American hunters, there were no grounds of either a legal or a customary nature for retaining control over access to the sporting experience. As the number of nonsubsistence hunters grew, the lines between these gentlemen and other hunters blurred even before they came into focus. By 1876, *Forest and Stream* was lamenting that "a quarter of a century ago the brotherhood of sportsmen, although not nearly so large, contained a greater majority of hardy, chivalrous, generous, and out-spoken gentlemen than at the present day."[21] And whereas the sporting gentlemen's knowledge of natural history was a *sine qua non*, Adam Bogardus, the American trap-shooting champion in the mid-1870s and a noted sportsman, was able to boast that

> I am no scientific naturalist, and what I know has not been derived from books. I cannot give the Latin names of birds or game, waterfowl, snipe, woodcock, etc., and if I could you would not care about them, because the constant repetition of them makes no impression upon the sportsman.[22]

Lewis epitomized the cast-aside image in his remarks concerning dress in the field:

> All those sportsmen whose occupation or profession makes it desirable that they should have white and smooth hands, and there are but few gentlemen whose employments do not require this, they ought, *ex necessitate rerum*, to wear gloves when shooting, as nothing to our eyes looks more *outré*, if not vulgar, than a coarse, scratched and scarred hand.[23]

By the 1870s coarse, scratched, and scarred hands would have been a sure mark of the successful outdoor adventure.

The corruption of English sport in America should be expected from features peculiar to the American experience. Wilderness skills were valued throughout the late nineteenth century, both in reverence to their past contribution to the struggle for independence and in anticipation of their contribution to the maintenance of the empire. Foremost among these skills was the handy use of weapons. The need to retain this skill justified not only the hunting activities of the sportsman but also the formation of rifle clubs and the construction of shooting ranges throughout the East.[24]

But this attribute of the sporting experience transfers completely to market and subsistence hunters who might be better able to muster the skills necessary for national defense. More to the point of distinguishing the market hunter from the sportsman was that for the sportsman, sport was a diversion and a renewal rather than a reflection of everyday existence. With

what one can forage from nature supplemented by a few provisions,

one will live well, grow hardy and tough as an Indian; lie down at
night on his fir-strewn couch, and sleep the sleep of tired, happy
childhood; and rise in the morning and take a dip in the clear flashing
water of some one of these lonely lakes, then settle around the spar-
kling fire, attack his breakfast with a vigor and an appetite previously
unknown. Such a life will do one good; the mental faculties keep pace
with the physical ones, and at the expiration of a two or three weeks'
trip to the woods, the tourist returns hardy in mind and muscle; large
and better hearted, ready to take his stand among men; ready to
advance an opinion and stick to it with a breezy, healthy earnestness
that will not be gainsayed.[25]

The growth in sport hunting was supported by the widespread popularity of outdoor recreation, the design of urban
parks and open spaces, and the devotion to physical culture
and diet. The Indianapolis area's "Kamping Klub" invited
members to withdraw "from the restraints of business and
society" and "revert into that indefinable state of existence,
whose joys and hardships make up the charm of camping
out,"[26] and Dr. Rothrock's Luzerne County, Pennsylvania,
school for the physical culture of boys promised to turn "pale,
petted and pampered objects of maternal solicitude" into out-
doorsmen "bronzed and ruddy with insatiable appetites."[27]
Although only the stodgiest New Englanders could see no
merit in sport for its own sake, unrelated as it was to the
business of life, *American Sportsman* nonetheless found the
need to justify recreational hunting and fishing as a reward for
the great material abundance achieved through past hard work
and diligence.[28]

Ultimately the American sportsman's efforts to distinguish
himself from the market or "pot" hunter could not rely on a
unique outcome: all hunters killed game. Much of the distinction came to rest on characteristics that were not outwardly
observable—attitude, motivation, and evaluation. In a major
sense, the sportsman was a hunter with a particular world
view.[29] He was also a hunter with a particular socioeconomic
status. The sportsman was generally urban, eastern, and wealthy

—characteristics that together were shared with few other hunters.

The sporting journals, in attracting advertisers, emphasized the wealth of their subscribers. The readers of *Forest and Stream* are "*purchasers*; they are men who can *afford to buy luxuries.*"[30] "They are not a miscellaneous class" nor part of the great masses who are "occupied with their avocation in their daily struggle for subsistence, and have no time or money for outdoor recreations and the gratification of natural tastes."[31] Similarly, the *American Sportsman* was claimed to go "among a class which is very free with money."[32] The kind of advertising found in these journals was directed at a readership much like that claimed by their editors. *Forest and Stream* enumerated some thirty-two distinct interests, which were represented in the journal's advertising pages, including books, disinfectants, horses, gymnasia, insurance, jewelry, soap, and steam yachts.[33] There was very little advertising for patent medicines and the like which filled the pages of popular magazines and newspapers of the day.

This influential class of hunters dominated the movement for restricted access to wild animals. Sportsmen sought closed seasons, controls on hunting methods, limitations on the size of daily or seasonal kills, and regulations on the sale, storage, and transportation of game. Some of these changes were meant to force universal compliance with practices which the true sportsman could claim had always guided his behavior; that they would incidentally reduce the profitability of marketing game was simply convenient. Other changes appeared to indicate the willingness of sportsmen to alter their own behavior in response to what they perceived to be the general interest. In the midst of these political maneuvers, sportsmen were jockeying for increased access to desirable wildlife populations through the purchase and lease of prime sporting grounds.[34]

The primary strategy by which sportsmen sought to improve their access to wildlife was the formation of local clubs nominally restricted to members like themselves. Among the first and most durable of these was the New York Sporting Association, which organized on May 20, 1844, under the motto *non nobis solum* (not for ourselves alone), "to protect sporting dogs from the oppressive laws then in force." The preamble to the constitution, adopted at the second meeting, suggests a considerably wider concern in the promotion and enforcement of laws to preserve "Game and certain varieties of Fish" and in promoting a "healthy public opinion in relation thereto."[35] In 1874, during the decade in which the protectionist spirit among sportsmen emerged to an important degree, the association, which had since become the New York Sportsman's Club, again changed its name to the New York Association for the Protection of Game.

This decade saw the creation of hundreds of clubs, similar in expressed purpose but differing widely in the extent to which they were interested or effective in modifying either behavior or the law.[36] These local clubs were most commonly called the (town or city) Rod and Gun or Sportsmen's Club, but many took on names in tribute to those with an early concern for wildlife, as did the Audubon clubs of Angola, Indiana, of Detroit, of Chicago, of Jacksonville, Indiana, of Johnson County, Iowa, and of Plymouth, Michigan, the Waltonian Club of New Haven, and the Forester clubs of Penn Yang, New York, of Rochester, of Buffalo, and of Beaver County, Pennsylvania. *Hallock's Club List* notes eight Forest and Stream clubs, and *Forest and Stream* listed the Hallock Sportsman's Association of Glens Falls, New York.

Most local clubs expressed at least sympathy with the spirit of game protection, but many in fact had as their primary purpose the construction and maintenance of a clubhouse or the rental of shooting lands. Some survived for decades, but most were short-lived. Some had restrictive membership policies, and some were open. Compare, for example, the South Side Sportsmen's Club of Long Island with the Kennebec (Maine) Association for the Protection of Fish and Game.[37] The former, incorporated in 1866, was limited to one hundred members, each of whom held a $500 share in the club and paid annual dues of $75. The constitution records that the club was established for the "protection, increase and capture of Game Birds and Fish, and for the promotion of Social intercourse among its members." The "Master of the Hounds" was charged with enforcing the club's hunting and fishing regulations, but the only such rule in the document consulted specified a daily limit of twelve trout. The club's interest in promoting wildlife protection beyond its own property appeared to be minimal. The Kennebec club, organized under the motto "He who causes two blades of grass to grow where only one grew before is a benefactor of the human race," had an unlimited membership, a one-time fifty-cent fee, and annual dues of twenty-five cents. The executive committee was directed to appoint a game warden for each town represented in the membership to enforce the game laws of the state. Each member was obligated to

solemnly promise, on my honor as a man and a citizen, that I will not violate the fish and game laws of this State myself, nor will I allow the same to be done if in my power to prevent it; that if any case of violation of said laws come under my own observation, I will at once report the same to the wardens of the Assocation; and that if I hear of any violation of said laws, I will take means to ascertain the truth of any such report, and if found true I will report to the wardens at once, that all such violations may be prosecuted.

Sportsmen did not limit their organizational efforts to cities and towns. Local clubs in New York joined into a state-wide association in 1858, although it was not well organized until 1865 and did not assume an important role in the coordinating of the interests of local clubs throughout the state until the 1870s. During this decade, state associations were organized in California, Connecticut, Delaware, Kansas, Kentucky, Illinois, Iowa, Maine, Maryland, Massachusetts, Michigan, Minnesota, Missouri, Nebraska, New Hampshire, Ohio, Pennsylvania, Rhode Island, Tennessee, Vermont, and Wisconsin. Many of these organizations were short-lived.[38] The key to the survival of others was the annual meeting which featured a pigeon shoot and other festivities. The gathering of the New York state association in 1881 included a ten-day shoot in which 20,000 birds were fired upon by 118 contestants for prizes worth $7,000. The opening meeting was a jovial affair. The assemblage was entertained by the Glee Club of the Washington Gun Club, which sang "a rollicking original ballad entitled 'The American Sportsmen's Song,'" and by William E. MacMaster, who read his original poem on field sports. "Its pertinent allusions were rapturously applauded."[39] In 1892, the state association of New York reorganized along lines more favorable to its named purpose, the protection of game.

Throughout the period 1870-1900, there were a number of attempts to organize multistate and, in particular, national sportsmen's associations as well as a single attempt to organize an international association of sportsmen. These were conceived largely for the purpose of generating interstate uniformity in the game laws, but in the absence of institutionalized procedures for adopting their recommendations, they were short-lived.[40]

THE MARKET HUNTER

Market hunters as a group were cast as the negative image of sportsmen. They were hunters who did not embody the qualities that sportsmen considered to distinguish themselves. They were farmers, frontiersmen, rural youth, and southern European immigrants. Because the deeply imbedded notions of egalitarian access to wildlife made explicit class distinctions unlikely to emerge, delineation between sportsmen and other hunters was usually based on behavior rather than on birth and social standing. Thus, sportsmen could responsibly make public examples of hide hunters who massacred bison for their skins and tongues, market gunners who threatened breeding populations by waterfowling in the spring, and rural settlers who killed exquisite and useful song birds for but a few feathers and a mouthful of meat.

These kinds of arguments carrried great weight in the public debate, but the preservation of the sportsman's identity was driven in part by a more basic concern for the erosion of a natural aristocracy which had, by inheritance, industry, and thrift, acquired economic and political power in a regime of private property.[41] And this was but the end product of a cultural evolution which brought humankind from savagery to barbarism to civilization. Charles Hallock observed that just as "nature is working for conciseness and unity" in eliminating species of life which do not "please [man's] fancy, aid him in subduing the earth, or possess economic value," so we "find the Caucasian race the dominant power of the world; the others being only its servants."[42] Among that leading race were those whose demonstrated skills in the business of life both guaranteed them a position of prominance and obligated them to husband the nation's wealth for the benefit of the less talented and fortunate. The game laws, written largely by sportsmen for the preservation of sporting opportunities, illustrated the exercise of this control. As wildlife, and especially game, became increasingly scarce, sportsmen were increasingly insistent in their claim not only to a manager's role but to outright ownership. *Forest and Stream*, in promoting its "No Sale Plank," first ennunciated in 1894, which would forbid the sale of game in all seasons, asserted "The Basic Principle" that "the game of this country belongs to the sportsman. . . . It is his, and he shall have it."[43]

For rural landowners and landless hunters, these assertions represented an expropriation of historic claims to resources. But, for sportsmen, a free-wheeling and open scramble for control of wildlife, perfectly symbolized by the market hunter, was now a threat to established patterns of control in the larger economy.

Underlying these arguments based on sound management was a more direct animosity based on ethnic and class lines. These views surfaced only occasionally in the early sporting literature, but, after the turn of the century, leading sportsmen, in promoting racial purity and inveighing against open immigration, were more likely to offer explicit class distinction. We find, for example, William Hornaday, chief taxidermist at the Smithsonian, thirty-year director of the New York Zoological Park, and highly influential opponent of market hunting, horrified by Italian immigrants who "root out the native American and take his place and his income. Toward wildlife the Italian laborer is a human mongoose. . . . The Italians are spreading, spreading, spreading. If you are without them today, tomorrow they will be around you."[44]

But even during the early period, it was the subsistence and market hunters, recognizing the political power and influence that upper-class sportsmen possessed and used to promote restrictions on access to game, who made direct references to the class distinction which separated groups of hunters. It was only wealthy sportsmen, they charged, who could respond to the decrease of game and its haunts by monopolizing prime hunting grounds behind walls of exclusive clubs, and it was only wealthy sportsmen who could afford to place the utilitarian value of game behind its sports value. On the latter distinction, sportsmen were plain wrong. Those who wasted game and did not look primarily to its utilitarian qualities were surely due less respect than those who insured its use as human sustenance.[45]

Beneath all of this rhetoric, there was indeed an indentifiable group whose members were primarily occupied in taking game and other wildlife as expediently as possible for sale and consumption. They worked for railroads and lumber companies; they hunted alone or as part of a small team; they delivered the game directly or through commission agents; and they knew the land and its resources. Some continued in the tradition of the trapper and mountain man of an earlier era, although the romanticism of that era was often lacking. For some, market hunting was a lifetime occupation—for others seasonal or short term. Entry into the ranks of the market hunter was easy. In the fields and forests, the only equipment required was already in the possession of the average rural household or was otherwise readily available. On the shore, where proficient hunters depended on battery boxes, "big" guns, and hundreds of decoys, start-up costs were higher. Whatever the technique, game dealers and their agents were always anxious to receive game offered for sale. While hunters around whom legends grew were few and far between, market hunting proved lucrative for numerous lesser-knowns. David W. Cartwright, in his *Natural History of Western Wild Animals, and Guide for Hunters, Trappers, and Sportsmen*, claimed that market hunting was ideal for the frontier farmer, as the five dollars per day that he could earn in that occupation was "certainly more than any neighbor of mine could pay me for working for him, even if he had work to be done."[46] Winter Brothers was paying its best Heron Lake, Minnesota, shooters five dollars per day plus room, board, powder and lead, but required at least one duck for each two shells.[47] But not everyone was suited for the trade. "Given a man of strength, of physical powers, courage, endurance, a close observation of the habits of animals hunted, skill in the methods used for their capture, ability to live in the woods without getting lost, a good trapping and camping outfit, and with all these a liking for the business, and the result is monetary success."[48] *Forest and Stream*, on the other hand, noted that "successful market hunting requires ability and pluck, which, if properly directed, would insure success in a more honorable pursuit."[49]

Market hunters were never a political force with which to be reckoned. Their very lifestyles militated against the collective action required to influence the course of political events. Few had or preserved conduits to centers of power, and any who stepped out of the back country to speak out against the slaughter of wildlife for the market would likely have been adopted by sportsmen for the purposes of interest-group affiliation.

Small teams of hunters working on salary or on commission for wholesalers represented the limits of organizational complexity. Their major task was the development of mechanisms for division of labor and of the receipts from the hunt. This is not to imply that hunters, as an interest group, had no stake in broader political decisions affecting the allocation of access to wildlife but rather that the structure of the "industry" precluded effective mobilization to achieve collective goals.

Conflicts demanding some form of resolution did arise as isolated hunters confronted one another over access to particular game populations or single animals. Browning claimed that he who drew the first blood had the right to the animal's hide "be it bear or deer."[50] This was substantiated by Cartwright, who added that "if a hunter has started up game, and it runs upon parties who may be close by, and they find it coming to them already wounded, they may kill it and take one half of the meat."[51] J. Lorenzo Werich reported that, in the region of the Kankakee River in Illinois, hunting parties divided the meat into as many piles as there were hunters:

One turns his back to the game and another points at each pile in turn and also asks whose it is. And the one with his back turned says who is entitled to the pile or bunch pointed at. But sometimes a heavy accent of [sic] signal by the one who points out is understood by the man whose back is turned. They sometimes give themselves the best pile of game.[52]

Hunting involves the active pursuit of animals; trapping involves their capture at a fixed point. Different rules were thus in order for trappers. In intensively hunted regions, territories could be claimed on a first-come, first-served basis, and they could be held by placing traps or setting up camp. A trapper understood that he could hang up another's traps if they were found within his territory.[53] Werich reports a much better defined arrangement in the region of his familiarity. The miles along the river were the base lines of claims which extended as far as profitable on either side of the water. "These claims were bought and sold almost the same as real estate and they were about as strong as the Clayton-Bulwort treaty."[54]

As market hunting became more lucrative and competition increased within a given region, the customary modes of interaction often gave way to a more ruthless pursuit of marketable game. It was this apparent change in behavior which drove Browning to a settled life at last. He reminisced that "all other hunters were not governed by the kind and fair dealing which used to regulate their actions in bygone years" and that "they began to take my traps, use them, and keep the game caught in them; thus greatly interfering with my sport."[55]

THE GAME DEALER

Unlike commercial hunters, those who sold game in urban markets were easily identifiable and in close proximity to one another. The volume of game sold was quite large through the turn of the century, as will be indicated in the next chapter.

Demand supported the largest markets in New York, Boston, Baltimore, and Philadelphia in the east and in St. Louis and Chicago in the west. Game dealers maintained stalls in the general produce and meat markets of their respective cities. The market appears to have been largely subject to competitive forces, though there is some indication that significant control was exercised by several large dealers during certain periods.[56]

Marketmen became increasingly better organized throughout the period of study, and they had significant impact on the course of game legislation in a number of states.[57] They were opposed to game laws favored by sportsmen only insofar as the means of enforcement appeared to discriminate against them. The legislative effort of marketmen was not therefore directed against the idea of protecting game during the breeding season but rather against restricting the sale of that game which was legally captured. Yet owing to the difficulty of enforcement in the field, the two restrictions generally occurred together. It seemed a clear and unwarranted restraint of trade to prohibit the sale of game on hand when the season for capture closed. The intent of the law did not require this prohibition, nor would it have stood the test of constitutionality if that were the intent. However, dealers believed that such restraints placed a significant burden on them and that they stood to gain from liberalized sale laws. As organizational costs were relatively small, there was incentive to act in concert to achieve this goal.

THE LANDOWNER

Wildlife exists on the land, and consequently changes in the landscape bring about changes in the patterns of wildlife that are a part of it. Landowners and others who modify that landscape by clearing, draining, cultivating, burning, mining, cementing, or abandoning play an intimate, if often indirect, role in the present history. The most pervasive cause of extinction and wildlife decline was not hunting for sport or market but the widespread changes in land use that accompanied growth and development.

During the early period of settlement, the landowner faced

few restrictions concerning land use. If anything, cultural attitudes toward the wilderness and institutional incentives that made final ownership dependent on improvements encouraged the disruption of natural environments. Patterns of settlement were defined by the rectangular survey; land grants and financial support stretched out the pathways of a transportation network; Indian "pacification" programs provided a safe expansion. The young nation was anxious to facilitate the transfer of the public domain into private property—a property that provided broad freedom of action with respect to management and resource extraction. Any alterations in wildlife populations which occurred as the result of these changes were merely fortuitous.

Although many rights of private property in land were clear, landowners' rights with respect to direct access to wildlife were not. The cultural norm suggested that wildlife, as common property, was available to all on a first-come, first-served basis, but the colonies, and later the states, had assumed the right to regulate access to wild animals. While it was true that landowners were not generally exempted from these regulations, it was also true that prosecution for violation was rare. In effect, he who occupied the land was free to take wild animals at his discretion; bounties on so-called noxious species promoted the hunt.

The lack of substance that characterized early wildlife legislation, while it freed the landowner from restrictions, also opened private lands to the uninvited who came in search of fish and game. Trespass laws were notably weak, especially governing vacant lands, for the common property nature of wildlife was often interpreted to apply irrespective of the particular location of the animals. The costs of trespass borne by landowners were small, however, owing to the small number of hunters and the general undervaluation of wildlife by landowners.[58]

During the period 1850-1900, there was little attempt to limit further the landowner's right to alter the landscape. On the contrary, the major land-disposal legislation of this period encouraged clearing, draining, and planting trees. The relationship between land-use patterns and wildlife abundance was not generally appreciated in the nineteenth century,[59] but, even if it were, this knowledge could not have overcome the driving force of expansion and growth which seemed to require unconstrained land transformation. There were a few sportsmen and naturalists who saw the interconnections on a small scale and who recommended plantings of wild rice and wild celery to encourage waterfowl, but wide recognition had to await twentieth-century game management.[60]

The developments during this period brought more significant changes in limiting the access of landowners to game on their own lands. The policy arguments for these limitations are persuasive insofar as valuable species migrate across property lines and insofar as landowners are otherwise incapable of capturing the gains from management. Although these conditions were generally in evidence, the argument was not made explicitly in such terms, and landowners perceived restrictive legislation largely as part of a general erosion of rural power. They were very protective of their historic liberty on the land, and they sought to retain it. Much easier to enforce than the private behavior of the landowner on his own property was the behavior of others on private lands, for the vigilant owner was present to detect trespass. As access to wildlife habitats became desired by increasingly greater numbers, landowners demonstrated a growing interest in securing increasingly strict trespass laws, both to protect their own land-based enterprises from damage and to regulate, often for profit, access to valued wildlife.

Although landowners only occasionally organized for the specific purpose of effecting change in the matter of their access to wildlife, they were often able to channel energy for these purposes through existing organizations and institutions. Agricultural societies, church and social clubs, and later the grange, provided obvious formats for the exercise of political power.[61] Local, regional, and even state governments were, in many areas, synonymous with agricultural and landholding interests. The virtues of private property in land were widely promoted by landowners and were the inspiration for models of game management that threatened the notions of common ownership on which both the sportsman and the market hunter depended.

NOTES

1. The development of American sport hunting and Herbert's role in that development are treated elsewhere. Recent sources include John F. Reiger, *American Sportsmen and the Origins of Conservation* (New York: Winchester Press, 1975), especially Chapter I, and James B. Trefethen, *An American Crusade for Wildlife* (New York: Winchester Press and The Boone and Crockett Club, 1975), Chapter 6.

2. Herbert was born in London in 1807, graduated from Cambridge in 1829, and went to New York in 1831, whereupon he became instructor of classics at the Reverend R. Townsend Huddart's Classical Institute. By the end of the decade, he had contributed to a variety of New York magazines and published several novels. In 1839, Herbert secured a position as a regular contributor to William Porter's *American Turf Register*. He was advised to select a pen name to protect his more serious business and literary pursuits. Porter rejected his first choice, Harry Archer (Harry because it was too familiar, and Archer because it was suggestive of obsolete hunting methods), and selected instead Frank Forester. His first important sporting work, *Field Sports in the United States, and the British Provinces of America*, 2 vols. (London: Richard Bentley, 1848), was derived from a series of articles that appeared in the *Democratic Review* in 1845-46. They were so well received that Herbert began to assume the character of Frank Forester more openly. "He appeared publicly in shooting-jacket and hunting-brogans, a fur cap on his head and a dog trailing at his heels," and he eschewed his literary friends. His life, in spite of outward appearances, was an unhappy one. In 1858, upon his second wife's suit for divorce following a three-month marriage, Herbert committed suicide. See Col. Thomas Picton's biography of Forester in David W. Judd, ed. *Life and Writings of Frank Forester*, 2 vols.(New York: Orange Judd Publishing Co., 1882), pp. 11-104. Picton "was the friend whom Herbert requested to share with him his last meal, and be present at the tragic ending" (ibid., p. 4.) Herbert's other sporting works include *Frank Forester's Fish and Fishing of the United States and British Provinces of North America* (New York: Stringer and Townsend, 1850), *American Game in Its Seasons* (New York: Charles Scribner,

1853), *The Complete Manual for Young Sportsmen* (New York: Stringer and Townsend, 1856), *Frank Forester's Horse and Horsemanship of the United States and British Provinces of North America* 2 vol. (New York: Stringer and Townsend, 1857), *Frank Forester's Sporting Scenes and Characters* (Philadelphia: T. B. Peterson, 1857), and *Fishing with Hook and Line: A Manual for Amateur Anglers* (New York: Published at the Brother Jonathan Office, 1858). For one of the earliest reports on the American sportsman, see T. Doughty, "Characteristics of a True Sportsman," which appeared in a three-volume collection of essays edited by T. and I. Doughty, *Cabinet of Natural History and American Rural Sports* (Philadelphia, 1830-35), and which is reprinted in John C. Phillips and Lewis Webb Hill, M.D., eds., *Classics of the American Shooting Field: A Mixed Bag for the Kindly Sportsman, 1783-1926* (Boston: Houghton Mifflin Co., 1930).

3. Herbert, *Field Sports*, I, 17-20.

4. Ibid., II, 334-36. James Lipton, in the belief that "our language ... deserves at least as much protection as our woodlands, streams and whooping cranes," has sought to restore respect for these "terms of venery." His discussion would have enriched considerably the vocabulary of the American sportsman [*An Exaltation of Larks* (New York: Grossman, 1968)].

5. E. J. Lewis, *Hints to Sportsmen* (Philadelphia: Lea and Blanchard, 1851), p. 16.

6. Ibid., pp. 16-17.

7. *American Sportsman* 3 (December 20, 1873), 184.

8. *Forest and Stream* 21 (November 1, 1883), 267.

9. H. H. Soulé, *Hints and Points for Sportsmen* (New York: Forest and Stream Publishing Co., 1889), p. 42.

10. Lewis, *Hints to Sportsmen*, p. 32.

11. *American Sportsman* 1 (September 1972), 4.

12. Herbert, *Field Sports*, I, 35-6.

13. Lewis, *Hints to Sportsmen*, p. 52.

14. John Dean Caton, *The Antelope and Deer of America: A Comprehensive Scientific Treatise upon the Natural History, including the Characteristics, Habits, Affinities, and Capacity for Domestication of the Antilocapra and Cervidae of North America*, 2d ed. (New York: Forest and Stream Publishing Company, 1881), p. 345.

15. Robert B. Roosevelt, *The Game Birds of the Coasts and Lakes of the Northern States of America* (New York: Carleton, 1862), pp. 271-72. By the time Robert Roosevelt's nephew, Theodore, gained influence in these definitional matters, it was the daring wilderness hunter and not the gentleman shooter of the eastern woodlands who epitomized

the American sportsman. See Chapter 3, text at notes 17-29 for a discussion of the evolution of the American sportsman through the turn of the century.

16. John Bumstead, *On the Wing: A Book for Sportsmen* (Boston: Fields, Osgood, and Co., 1869), p. 112.

17. Lewis, *Hints to Sportsmen*, p. 53.

18. *Forest and Stream*, 8 (April 5, 1877), 132.

19. The journal's full title suggests a broad concern: *Forest and Stream, a Weekly Journal Devoted to Field and Aquatic Sports, Practical and Natural History, Fish Culture, the Protection of Game, Preservation of Forests, and the Inculcation in Men and Women of a Healthy Interest in Outdoor Recreation and Study.* Hallock's previous journalistic experience included a year as assistant editor of the New Haven *Register* (1855-56) and several years on the staff of the New York *Journal of Commerce* of which his father was editor. In 1871, he was among the founders of the Blooming Grove Park Association in Pennsylvania, and served as its secretary for two years (see Chapter 4, text at notes 25-28). In 1879, Hallock left *Forest and Stream* for reasons that included, according to his successor George Bird Grinnell, fiscal mismanagement and excessive drinking (see Chapter 6, note 35). Throughout the remainder of the century Hallock concerned himself with the fate of wildlife resources, and was particularly outspoken on the matter of uniform hunting seasons (see Chapter 6, text at notes 17-21). He served short terms as the coeditor of *Wildwood's Magazine* (1888) and of *Western Field and Stream* (1896). Other important sporting journals were Nicholas Rowe's *American Field* (Chicago), The Parker Brothers' *American Sportsman*, later *Rod and Gun and American Sportsman* (West Meriden, Conn., and New York, 1871-77), and G. O. Shields's *Recreation* (New York, 1894-1905). An 1889 survey of American sporting periodical literature included fifty-six journals covering a wide variety of sports [*Recreation* (*Wildwood's Magazine*) 3 (June 1889), 81-82]. See also Frank Luther Mott, *A History of American Magazines*, 5 vols. (Cambridge, Mass.: 1938-68).

20. Kennebec Association for the Protection of Game, "A Catalogue of Officers and Members, Act of Incorporation, Constitution, By-Laws and Obligations, and President's Address, and Fish and Game Laws" (1874), p. 11.

Circulation and distribution of the major journals is approximated as follows: *American Sportsman* guaranteed its advertisers 15,000 readers for the first three months [1 (November 1871), 7] and maintained a subscription list of about 10,000 in 1875 [5 (March 6, 1875), 360]. *Forest and Stream* mailed about the same number of copies according to its own report of 1879 [11 (January 2, 1879), 446], and they were delivered to some 2,400 post offices in the United States and Canada and to 100 more in thirty foreign countries [11 (December 12, 1878), 386]. An earlier report showed a total of 1726 post offices in the United States— 300 in New York, 701 elsewhere in the east, and 725 in the west [10 (April 11, 1878), 182].

21. *Forest and Stream* 5 (February 3, 1876), 409.

22. Adam H. Bogardus, *Field, Cover, and Trap Shooting*, 2d ed. (New York: published by the author, 1878), p. 17. Bogardus, not known for modesty, summarized his early life: "I was born in Albany County, New York, and began to shoot at fifteen years of age. I was then a tall, strong lad, and have since grown into a large, powerful, sinewy, and muscular man" (ibid.).

23. Lewis, *Hints to Sportsmen*, p. 246. It is remarks of this kind which urge caution in accepting broad claims of sport as the great equalizer, as for example, the same author's title-page dictum: "In the city, people of different ranks stand scowling and apart; but when they go to hunt, to fish, or to any other sport of occupation in the fields, they are fellows. Nature thus makes brotherhood; and if all mankind would study Nature, all mankind would be brothers."

24. In particular, the National Rifle Association was organized in 1871; two years later a practice range was established at Creedmoor, on Long Island. A number of international shooting matches followed.

The relationship between familiarity with the rifle and military prowess had no major test until 1898. *Field and Stream* jumped at the chance to explain the American victory over the Spanish on the basis of its sporting tradition. "All young America, practically, are sportsmen. As soon as a boy is big enough and strong enough to bear a gun, he finds business to do, at least for a few weeks of each year in the fall, along the sloughs and fields and marshes of his native State or some adjoining one. . . . He is a campaigner, with good, tough, muscular legs, no nerves, a quick eye and a sudden trigger finger, before he has reached the second book of Euclid or can tell a Latin verb from the Hellespont. He can make camp, build a fire and do his own cooking. He is a soldier in the rough before he knows enough to stand at attention, and all that has to be done is to smooth him off, like a clay model, teach him to march in step, and send him out to give any casual belligerent points on hitting bull's eyes" [3 (July 1898), 230].

25. *Fur, Fin, and Feather: A Compilation of the Game Laws of the Different States and Provinces of the United States and Canada; to which is added a list of Hunting and Fishing localities, and other useful information for Gunners and Anglers*, revised and corrected for 1871-72. (New York:

M. B. Brown and Co., 1871), p. 194. This mixture of the utilitarian and the romantic is further in evidence in the editorial announcement of *Forest and Stream*, which begs the reader to return to "first principles." "Remove temporarily our modern appliances and we are helpless. Let us acquire the rudiments anew. We know not at what moment the storm may lay us ashore upon an island uninhabited, . . ."

26. *Forest and Stream* 12 (July 10, 1879), 450.

27. *American Sportsman* 8 (August 5, 1876), 298-99.

28. *American Sportsman* 2 (May 1873), 121. The characteristics of the ideal sportsman during this period were decidedly masculine. Yet it was hard to deny to women the health-generating effects of sporting activities as these were increasingly emphasized. *Forest and Stream*, in an early discussion of the matter, concluded the following: "Although opposed to 'woman's rights' in an Anthonian or Walkerian sense, we are willing champions for her rights to health and happiness: . . . After a hard day's work at those heroic sports in which she cannot partici-pate, how pleasant to return and find that brightest ornament of the home, whether wife, sister or mother, waiting to receive you?" The journal went on to admit that there was indeed a place for women in field sports, as for example, in fly casting, where a "skillful hand and supple wrist" are necessary for success [12 (May 8, 1879), 270].

29. Reiger has also found the concept of "world view" useful in characterizing the American sportsman (*American Sportmen and the Origins of Conservation*, p. 29).

30. *Forest and Stream* 5 (January 16, 1876), 345. The hunt itself was an expensive endeavor when conducted in the proper manner. The eastern sportsman may have made his purchases from W. Holberton & Co. of New York which sold "such articles as enter into and are necessary to the outfit of gentlemen visiting the Hunting and Fishing Grounds of the United States and the Canadas." In a year in which the average earnings of nonfarm employees was $400, he may have pur-chased a tent, 9' x 9' with a 4' wall ($25), a jack lamp and support, for night-hunting ($9.25), a two-man boat ($40), a folding cot ($8), a corduroy shooting suit ($28; $40 in velveteen), hunting boots ($12), a breech-loading shot gun ($50-$300) and cartridges ($37-$55 per 1000, depending on calibre), a Sharp's rifle ($25-$300) and bullets ($8.75-$15 per 1,000, depending on calibre), wooden duck decoys ($10.50 for teal and $12 for mallard or canvas-back), and Newhouse traps with chains (#0 for rats, $4 per dozen; #4 for beaver, $16.50 per dozen; #6 for grizzly bears, $280 per dozen) [W. Holberton & Co., *Illustrated Cata-logue and Hand-book for Sportsmen* (New York, 1878)].

31. *Forest and Stream* 2 (March 12, 1874), 72.

32. *American Sportsman* 1 (November 1871), 7.

33. *Forest and Stream* 8 (April 5, 1877), 132.

34. See Chapters 4 and 6 for a detailed discussion of these efforts.

35. "Historical Sketch," *Constitution and List of Officers and Members of the New York Association for the Protection of Game*, 1891, and George Bird Grinnell, "American Game Protection, A Sketch," in George Bird Grinnell and Charles Sheldon, eds., *Hunting and Conservation* (New Haven: Yale University Press, 1925). On early fishing clubs, see Charles Eliot Goodspeed, *Angling in America: Its Early History and Literature* (Boston: Houghton Mifflin Company, 1939).

36. *Hallock's American Club List and Sportsman's Glossary*, published in 1878 (New York: Forest and Stream Publishing Company), listed 316 clubs concerned primarily with hunting game animals, another 237 concerned primarily with riflery, 186 with trap-shooting, 36 with angling, and 18 with fox hunting or coursing; this writer's casual observation noted 390 local sportsmen's clubs, formed before 1881, in the pages of *Forest and Stream* and *American Sportsman*; finally, Will Wildwood's *Sportsman's Directory*, published in 1891 (Milwaukee: Fred E. Pond), provided a list of some 950 local clubs in thirty states and territories. [Will Wildwood was the pen name of Frederick Eugene Pond. Pond (1856-1925) was born in Wisconsin, helped to organize the State Sportsman's Association there in 1874, edited *Turf, Field and Farm* from 1881 to 1886, began his own sporting journal (*Wildwood's Magazine*) in 1888, and served as the secretary of the National Game, Bird, and Fish Protective Association from 1893 to 1895.] See also Reiger, *American Sportsmen and the Origins of Conservation*, pp. 39-40.

37. The "Constitution and By-Laws" (1874) and "A Catalogue of the Officers and Members, Act of Incorporation, Constitution, By-Laws and Obligations, and President's Address, and Fish and Game Laws" (1878) of these clubs, respectively, are contained in the Beinecke Rare Book and Manuscript Library of Yale University. See also Gherardi Davis, *The Southside Sportsmen's Club of Long Island* (privately printed, 1909), which records that the club owned 2,324 acres and leased an additional 1,147 acres. A total of 416 different members had belonged since the club's founding. Goodspeed reports that early members of the club included Andrew Carnegie, restauranteur Lorenzo Delmonico, mayor Abram Hewitt, and George Bird Grinnell (*Angling in America*, pp. 190-91).

38. In fact, Wildwood's *Directory* indicates that, in 1891, state associations existed only in California, Connecticut, Delaware, Illinois,

Indiana, Massachusetts, Michigan, Missouri, Nebraska, New York, Oregon, Texas, Vermont, and Virginia (pp. 9-10).

39. *The New York Times* (June 21, 1881), p. 5. The paper provided daily coverage of the events of the convention for the ten days beginning June 20.

40. The proposals of these associations will be considered in more detail in Chapter 6, text at notes 22-34.

41. See Ralph Henry Gabriel, *The Course of American Democratic Thought: An Intellectual History Since 1815* (New York: The Ronald Press, 1940), Chapters XIII and XIV. Reiger suggests that gentleman sportsmen (at least Hallock and Grinnell) defended proper styles in sport in reaction to the apparent surrender of a noncommercial aristocracy to the "multi-faceted trend toward commercialism and Philistinism that was accompanying the rapid industrialization of American society" (*American Sportsmen and the Origins of Conservation*, pp. 32-34).

42. *Forest and Stream* 3 (December 3, 1874), 264. The glossary in Hallock's *Sportsman's Directory* (1879) states that an "essence peddlar" is a skunk; a "Sportsman" is one "who hunts or fishes for pleasure, in a legitimate and scientific manner, without regard to pecuniary profit"; and a "Greaser" is "A Mexican, so named for his greasy appearance" (pp. 701-13).

43. *Forest and Stream* 57 (November 23, 1901), 401. See Chapter 4, text at notes 3-7, for a summary of the arguments surrounding the ownership of wildlife.

44. William Hornaday, *Our Vanishing Wildlife: Its Extermination and Preservation* (New York: The New York Zoological Society, 1913), pp. 101-2. Madison Grant, another leading sportsman, member of the Boone and Crockett Club, executive officer of the New York Zoological Society from its inception in 1896 to 1925, and principle organizer of the American Bison Society, the National Park Association of Washington, and the Save the Redwoods League, was also a leader in the eugenics and anti-immigration movements. He authored *The Passing of the Great Race* in 1916 and played an important role in securing the 1924 legislation restricting immigration. (*National Cyclopedia of American Biography*, vol. 29, p. 319, and *Dictionary of American Biography*, Supplement 2, p. 256.)

45. Clear distinctions were hard to make. The major sporting journals that promoted the image of the sportsman and that claimed to represent him were certainly read by large numbers of hunters whose own style did not fit the model. In fact, much of this abstract debate

was played out on the pages of these journals. They contained information useful to all hunters: exploits of hunters which revealed abundance and diversity of game, urban game prices, changes in the law, and natural history, to mention but a few.

46. David W. Cartwright, *Natural History of Western Wild Animals, and Guide for Hunters, Trappers, and Sportsmen* (Toldeo: Blade Printing and Paper Co., 1875), p. 5. Many remarks on the financial success of market hunters which occur in the literature are clouded by a critical view of their accompanying life style and of their effects on the stock of game animals. Dodge, for example, observed that "in season or out of season they kill everything that comes in their way. If the animal is unfit for food its skin may bring a dime or two. Once in two or three months they will go to the nearest railroad town, sell off the peltries they have accumulated, buy a little flour and bacon, a bag of salt, and a few beans. The balance of the money is either lost at a faro bank or spent in a roaring spree, after which they return to the wilderness. These men think only of to-day. The game have no respite or opportunity for recuperation, and must soon disappear" [Richard Irving Dodge, *The Plains of the Great West and Their Inhabitants* (New York: G. P. Putnam's Sons, 1877), p. 117].

47. David and Jim Kimball, *The Market Hunter* (Minneapolis: Dillon Press, 1969), p. 51. "Although the Heron Lake market hunters were wing shooters, if a hunter found himself below the fifty percent requirement at the end of the day he could easily raise his average with a few pot shots at sitting flocks."

48. Cartwright, *Natural History*, p. 5.

49. *Forest and Stream* 21 (November 8, 1883), 281.

50. Meshach Browning, *Forty-four Years of the Life of a Hunter* (Philadelphia: J. B. Lippincott & Co., 1864), p. 58.

51. Cartwright, *Natural History*, pp. 4-5.

52. J. Lorenzo Werich, *Pioneer Hunters of the Kankakee* (printed by the author, 1920), pp. 50-51.

53. Cartwright, *Natural History*, pp. 4-5.

54. Werich, *Pioneer Hunters*, p. 87.

55. Browning, *Forty-four Years*, pp. 358-59.

56. One of these influential dealers was Amos Robbins of the Manhattan firm of A & E Robbins. H. Clay Merritt, through whose hands passed a significant share of midwestern game on its way to market, claimed that "his sway was absolute. All the managers of restaurants and hotels looked to him for their game. When he told them to drop one kind, it was done. When he told them to take up

another it was done without asking why. . . . The moment he was dead [c. 1890] the game business was palsied, and the prices which for long years waited on his footsteps were rudely cast aside" [H. Clay Merritt, *The Shadow of a Gun* (Chicago: F. T. Peterson Company, 1904), pp. 221-23].

57. This was particularly true in Illinois, beginning with the significant enforcement of the game laws in the late 1880s. Merritt remarked: "No one but the dealers in Chicago took the trouble and expense to go to Springfield and the law makers there passed whatever was asked by three or four dealers carte blanche" (*Shadow of a Gun*, p. 306).

58. See Chapters 3 and 4 for a discussion of the increase in the knowledge of farmers and other landowners.

59. One notable exception was George Perkins Marsh's *Man and Nature*, published in 1864 and revised in 1874 as *The Earth as Modified by Human Action*.

60. See in particular Aldo Leopold, *Game Management* (New York: Charles Scribner's Sons, 1933).

61. Lance E. Davis and Douglass C. North, in *Institutional Change and American Economic Growth* (Cambridge: At the University Press, 1971), refer on several occasions to the fact that the existence of the Grange allowed farmers to engage in relatively costless forms of collective activity as their organizational costs had largely been met.

3

WILDLIFE IN DECLINE

> Once, the sky used to be black with ducks everywhere you looked. There appeared to be no end to their numbers. What a shame to see them go.
>
> market gunner, with an estimated lifetime kill of 500,000 ducks.

The white-tailed deer was common, if never plentiful, in Vermont before white settlement. The extensive hunting and clearing that accompanied early growth drastically reduced deer numbers, and by 1860, the species was nearly extinct in the state. In 1865, the legislature enacted a ten-year closed term for deer. Two years of hunting followed the expiration of this ban, but no records were kept that might indicate the abundance of the species. In 1878, members and friends of the Rutland County Deer Association acquired seventeen deer from private herds in New York State and released them in the central Vermont woods. "It was understood that years must elapse—many of them—before the plan could be realized, and not one of [the sponsors] ever expected to live to kill a deer in Vermont; it was an investment for posterity."[1] The legislature restored the closed term and renewed it at intervals through 1900. The members of the Rutland association volunteered to enforce the ban to the best of their abilities.

By the mid-1890s, deer numbers had increased. In 1894, the Vermont commissioners of fish and game observed that evidence against violators of the closed term was becoming difficult to secure as "nearly all the inhabitants in the more sparsely settled communities have tasted venison illegally killed."[2]

Residents of the northern counties made it clear that they would actively prevent illegal hunting in exchange for a short open season. In 1897, under the pressure of complaints of deer damage to crops and gardens, the legislature terminated the closed term and permitted buck hunting during the month of October. The press had led hunters to believe that the "ravages of deer resembled the grasshopper plague of the west," and out-of-staters poured into Vermont in anticipation.[3] Only 103 deer were reported killed during that month. The commissioners claimed that enforcement was exceptional, and that "if anyone killed a doe it was not carried out of the woods."[4]

There has been an open deer season in Vermont every year since 1897. The legal harvest increased steadily to 2,644 in 1911, fluctuated below that figure through the late 1930s, and grew rapidly to a peak of 17,384 in 1966.[5] In 1974, an estimated minimum deer population of 155,000 generated a total buck and bow season harvest of 12,834.[6]

The lessons of this story should be carefully drawn. Hunting is not of singular importance in determining the size and health of wildlife populations. The growth of deer herds in Vermont was paralleled by the abandonment of fields and farms which restored much needed winter range. The white-tailed deer is a resilient and adaptable species. It has recovered a substantial portion of its primitive range and invaded new regions. It has become a pest in some areas and has destroyed its own habitat in others. Other species have fared differently. Some—like the passenger pigeon and the Carolina parakeet, for example—were destroyed with the settlement of the continent. Others, such as the turkey and the prairie chicken, were locally extirpated. Species differ in their adaptability to change. Those requiring a specialized habitat cannot survive elsewhere even in the absence of hunting pressure, yet in the appropriate habitat, they may sustain a considerable hunting intensity without appreciable decline.

Despite local fluctuations, wildlife populations were generally in decline throughout the nineteenth century. As the century wore on, the declines became more pronounced and signified to the sporting and scientific communities that the customary free access to wildlife could no longer continue. The collective response to wildlife scarcity owes its origins in part to broad social and institutional changes of the late nineteenth and early twentieth centuries, and in part to the observed patterns of scarcity. This chapter surveys the fate of wildlife at the hands of hunters and land users.

THE SPORT HUNT

The impact of sport hunting on wildlife populations grew with the number of sportsmen and the spread of their expeditions. Eastern wildlife was reasonably accessible to the sportsman even before the railroad, but western populations were available only to the hardiest travelers. Poor transportation was a minor concern where the risk of Indian attack loomed large, but safety and hunting moved west with the railroads.[7] By the 1880s, transportation companies were actively attracting sportsmen to regions along their routes. Brochures and pamphlets proliferated, detailing game to be found, accommodations along the route, general prices, and even lists of reputable guides whom the sportsman might hire. The pamphlets were written as if each railroad had been routed with particular care to pass through famous and well-stocked game haunts. "To attempt to specify the localities where good shooting may be had would simply be to call the roll of the stations along the line of the Great Northern road."[8]

Other sources available to the sportsman were general guides to the hunting and fishing grounds of the nation. Typical is Charles Hallock's *Sportsman's Gazeteer and General Guide*. The fifth edition (1879) provides over 200 pages of listings by state and county. Anaheim, California, not now known principally as a haunt of wild game, was described by these remarks, which are illustrative of the information provided: "Deer, rabbits, hares, quail, geese, ducks. An occasional grizzly bear is shot within thirty or forty miles. Reached by branch of Southern Pacific Railroad. Board $2 to 2.50."[9]

Among the more popular eastern sporting grounds was the Adirondacks of New York. Though well known by trappers and a few adventuresome sportsmen in the early nineteenth century, the region was largely untraveled in 1858 when Ralph Waldo Emerson, Louis Agassiz, James Russell Lowell, W. J. Stillman, and company set out for what would become known as the Philosophers' Camp near Long Lake.[10] But by the 1870s, in part due to the wide distribution of William H. H. Murray's *Adventures in the Wilderness; or, Camp Life in the Adirondacks* (1869), the Adirondacks attracted crowds of sportsmen and tourists. John B. Bachelder, in his *Popular Resorts and How to Reach Them*, found the region, in 1874, a desirable place for escape from city life. "At present the Adirondack may boast of its primitive charms; but the region will, doubtless, be materially altered, in this respect, ere long, as visitors to this region are annually numbered in the thousands."[11] For others, the area had already lost its virtues. Wilder regions were superior to the "Adirondacks, where, I believe a man can not bathe in a mountain lake without cutting his feet on the remnants of some broken whiskey bottle or lie down at night without staining his blanket on a cigar stump—eloquent traces of some of our modern woodsmen."[12]

The same problems plaguing the twentieth-century wilderness seeker plagued the sportsman of the day. Favored locations became popular and lost the advantages which initially recommended them. In some states, special measures were adopted to control the onslaught of hunters. Maine, for example, required all sportsmen who were not residents to employ licensed guides.[13] By the late 1890s, the Maine guide system had achieved popularity as far west as Colorado. Many sportsmen, however, shared the sentiments of Emerson Hough, western correspondent to *Forest and Stream*, who wrote from Colorado:

... We have been proud of the fact that we had a region where a man could get lost if he wanted to, and where perhaps he could not get a guide if he cared to do so. We have felt that wild sport meant wild country, country uncharted and untabulated, country which had its own secrets and its own ways. It is this sort of country which really appeals to the man who has a hunter's instinct of getting away from business places and business methods. . . . The deer killed under the tutelage of a licensed guide will never have the same value as that killed by the sportsman himself in a country which he discovers for himself, and which he loves because he fancies it for the time his own.[14]

The popular eastern hunting grounds might provide recreation but not true sport. The "hotel guides," kept on retainer by resorts, emerged to service a class of tourists interested only in "idling away their time in the beaten round of summer hotel life."[15] The charge that massive Adirondack deer killings resulted from the sport of tourists merited little serious consideration. "Slaughter by sportsmen! Faugh! There's not one in a hundred who could catch a deer without the aid of his guide, even if he were at the point of starvation."[16] Presumably all of the able sportsmen were in Colorado.

The abundance and distribution of wildlife depends not only on the abundance and distribution of hunters but also on the choice of species that form the object of the hunt. The gentleman sportsman favored small, upland species, that required some skill in capture. Thomas Alexander claimed that "upland shooting is the very essence and marrow of Nimrodism."[17] In particular, the woodcock was considered the "ne plus ultra of all shooting on the wing."[18] Lewis was unable to explain why he and other sportsmen had this view about the woodcock, "but we know that we have it."[19]

As hunters increased in number and the haunts of highly regarded species became crowded, sportsmen modified the list of species deemed worthy of pursuit. Part of this change was a direct result of the move west. Species not part of the English heritage, such as the antelope, mountain sheep, and grizzly bear, were readily accepted in the American sportsman's vocabulary. The Wild West became the locale of the "wilderness hunter" as portrayed by Richard Irving Dodge in *Plains of the Great West* and by Theodore Roosevelt in *The Wilderness Hunter* and other works. But in the west as elsewhere, species favored by the sportsman provided him with a challenge and required skills beyond those which anyone could acquire by the mere possession of a gun. Thus, for

Dodge, the sport gained in the chase after one black-tailed buck was "more full of pure satisfaction . . . than the murder of an acre of buffalo."[20]

But as desirable game became scarce, sportsmen wondered about their future targets:

The prospect for the sportsman of the future is indeed gloomy, unless he shall make game of the pests and become a hunter of skunks and a shooter of crows and sparrows. Who can say that a hundred years hence the leading sportsmen of the period will not be wrangling over the points and merits of their skunk and woodchuck dogs and bragging of their bags of crows and sparrows?[21]

Species such as the alligator, grizzly bear, and mountain lion, once considered "vermin and outlaws," were "transformed into game animals and elevated to the rank of those deserving the protection of close seasons and restrictions as to modes of capture."[22] George O. Shields's compilation, *The Big Game of North America*, published in 1890, includes an essay entitled "Alligator Shooting in Florida." "The sight of the huge, glittering body, as it lies basking in the sunshine, may well cause [the sportsman's] heart to beat as hard and his breath to come as heavy as though a more beautiful game animal lay before him."[23]

The designation of new game species was accompanied by the development of new hunting technologies. The important changes were the replacement of the flint-lock with percussion caps and the replacement of the muzzle-loader with the breech-loader. But new technologies only slowly pushed aside the old. The breech-loader, developed in France in the 1840s, outsold the muzzle-loader five to one in England by the 1860s, but "only some twenty or thirty" were in use among American sportsmen at that time.[24] By 1873, the breech-loader represented about one-fourth of sales and, by 1883, about three-fourths.[25] Other weapons, such as punt or swivel guns, mounted on boats as small cannons and firing huge charges, were developed primarily by market hunters. While sportsmen opposed market-hunting methods by definition, a few questioned their own simple weapons.[26] The objection was based on the belief that while technology continually improved the sportsman's abilities, "the luckless wild things are endowed no better than when a bow and arrow were their worst foes. . . . For my part I should be quite contented to see the total abolition of what is known as sport with a gun."[27]

Many of those who turned away from shooting turned toward photography. This shift was facilitated by refinements in photographic method which allowed work in the field. George Shiras III, who developed a technique for night photography in which animals would simultaneously trip the flash and the shutter and who would later introduce the first legislation seeking federal control of migratory birds, was important in spreading enthusiasm for photographing wildlife. "Every camera hunter must admit that more immediate and lasting pleasure is afforded in raking a running deer from stem to stern, at twenty yards, with his 5 x 7 bore camera than driving an ounce ball through its heart at 100 yards."[28] Conveniently, photography required more of the qualities which sportsmen had always considered to characterize themselves. "Does it not require a greater amount of courage to face an angry mountain lion with a camera than with a Winchester .30-.40?"[29] Moreover, the sportsman-photographer was limited neither by closed seasons nor by other hunting restrictions. To encourage the use of cameras by sportsmen, *Forest and Stream* ran a photography contest in 1892 and published the winning pictures. Other sportsman's journals—in particular, G. O. Shields's *Recreation*—gave considerable attention to the activity. Around the turn of the century, the sportsman could have found guidance and encouragement in Chapman's *Bird Studies with a Camera*, Dugmore's *Nature and the Camera*, Brownell's *Photography for the Sportsman Naturalist*, and Thomas's *Hunting Big Game with Gun and Kodak*.

THE MARKET HUNT

It may be difficult for the twentieth-century urbanite, purchasing at retail a limited variety of meat and fowl and participating, perhaps, in the fall hunt for a single deer, to recognize the omipresence of wildlife in nineteenth-century American diet and fashion. Virtually all species of birds and large mammals were marketed in cities around the nation. Thomas F. De Voe, in *The Market Assistant*, describes those which he claims were available, albeit rarely in some cases, in the public markets of New York, Brooklyn, Boston, and Philadelphia during the 1860s. The birds included were two species of swan, four of geese, twenty-six of duck, the horned grebe, coot, loon, wild turkey, quail, woodcock, passenger pigeon, six species of grouse, forty-three of marsh and shore birds including snipe, rail, cranes, gulls, and curlews, and numerous small song and insectivorous birds. The mammals described were the bison, white-tailed deer, elk, caribou, moose, mule, deer, antelope, big-horn sheep, mountain goat, hare, rabbit, squirrel, black bear, raccoon, wild cat, oppossum, woodchuck, porcupine, skunk, beaver, otter, badger, and muskrat.[30]

The game supper, still popular in many rural communities, offers another view of consumption patterns. Among the more famous was the annual Thanksgiving meal at Drake's Hotel in Chicago. The offerings for the 1886 feast included white-tailed deer (soup, boiled tongue, roast saddle, broiled steak, cutlet), black-tailed deer (roast), mountain sheep (boiled leg, roast), black bear (boiled ham, roast, ragout "Hunter style"), cinnamon bear (roast), buffalo (boiled tongue, roast loin, broiled steak), antelope (steak), elk (roast leg), oppossum (roast), raccoon (roast), rabbit (broiled, "Braise, Cream sauce"), jack rabbit (roast), English hare (roast), goose (roast), redhead duck (roast), canvasback duck (roast), bluewing teal (roast, broiled), widgeon (roast), mallard (roast, broiled), wood duck (roast), butterballs (broiled), wild turkey (roast), quail (roast), pheasant (roast, broiled), prairie chicken (roast, "en Socle," filet "with Truffles," salad), spotted grouse (roast), sandhill crane (roast), jacksnipe (broiled), English snipe (broiled), blackbirds (broiled), reed birds (broiled), trout (broiled), and black bass (baked), as well as several "ornamental dishes" such as "Red-Wing Starling on Tree," "The Coon out at Night," "Pyramid of

Game en Bellevue," and "Pyramid of Wild-Goose Liver in Jelly."[31]

The specific impact of the market for wildlife on the distribution and abundance of animal populations is difficult to assess. For several species, most notably the passenger pigeon and the bison, a fairly complete and accurate history is available owing to the rapidity and density of the hunt and the publicity it received. These histories will be recounted briefly below. For other important species, as the white-tailed deer and game birds, only isolated marketing reports are in evidence. Most often these reports included either price or quantity data, which makes it difficult, in retrospect, to associate market movements with changes in demand and supply. An occasional record offers both. In 1873, an estimated $500,000 of game was sold in the Chicago retail markets. This total included 50,000 dozen prairie chickens at $3.25 per dozen, 25,000 dozen quail at $1.25 per dozen, 5,000 dozen pigeons at $1.00 per dozen, 30,000 pounds of elk, 400,000 pounds of buffalo, 450,000 pounds of venison, 10,000 pounds of bear, and 225,000 pounds of antelope at five, seven, eight, eight, and ten cents per pound, respectively.[32] In the retail markets of Los Angeles and San Francisco in the 1895-96 season, 501,171 game birds brought $62,000, including 177,366 quail, 82,525 teal, and 385 crane at average prices of nine, ten, and fifty cents each.[33]

Isolated price reports for numerous market areas appear throughout the literature. *Forest and Stream* printed weekly price quotations for the New York retail market from 1874 to 1876 and occasional quotations in other years. These prices, for the week of November 19, 1874, suggest the relative value of game in the urban east: venison, 25 cents per pound; quail, $3.75 per dozen; pigeon, $3.00 per dozen (stall fed) and $2.25 per dozen (wild); redhead ducks, $1.50 per pair; canvasback ducks, $2.50 per pair; ruffed grouse, $1.25 per pair; and robins, $1.00 per dozen.[34] Before refrigerated storage and transportation, prices in western markets were considerably lower. In December 1874, game in St. Louis was "actually cheaper than the most ordinary foods." Venison could be purchased at three cents per pound, ducks at $1.50-2.00 per dozen, and grouse at a dollar per dozen.[35]

Though the traffic in beaver pelts had slowed to a trickle, there remained an active market for these and other furs throughout the century. Prices varied widely according to size, coloration, and quality as well as local market conditions. In New York, beaver skins, which had brought no more than $3.00 in 1872, brought up to $6.50 in 1879. Conversely, mink, which had brought up to $5.00 in 1872, brought no more than $1.50 in 1879. Other furs brought the following prices in 1879 depending on quality: bear, $5.00-12.00; black house cat, $.25-.50; silver fox, $5.00-50.00; grey fox, $.30-.40; muskrat, $.04-.15; oppossum, $.06-.10; raccoon, $.15-.65; rabbit, $.02-.03; skunk, $.60-1.00; and wolf, $.75-2.50.[36]

Reports of quantities are also scattered throughout the literature, often in the form of self-proclaimed tributes to hunting prowess. A single hunter in California claimed to have brought to market 6,380 geese, 5,956 ducks, 367 sandhill cranes, and 60 swans during a nine-month period beginning September 1877, and another shooter in Iowa reported a personal bag of 2,000 prairie chickens, 100 woodcock, and "two or three wagon loads of ducks," among other game, during the 1873 "season."[37] Although these spectacular hunts may seem necessarily detrimental to the populations in question, no clear judgment can be made in the absence of information on population size and condition. But whatever their impact on populations, these hunts were held up by the sporting press as examples of irresponsible behavior that might be curtailed by increased restrictions on the market gunner. It seemed not to matter that, as often as not, these individual harvests were undertaken by hunters who would wish to be included among the ranks of the sportsman.

Other quantity reports referred to the total numbers of animals taken in particular states or regions. Early reports are of doubtful accuracy, but with improvements in record keeping by common carriers and with the development of hunter licensing and game tagging in the latter part of the century, they began to offer reliable information on the condition and size of wildlife populations. The general decline of the white-tailed deer herd is forcefully indicated by estimates which show the Michigan harvest of 60,000 in 1880-81 was matched in 1910 by the harvest in all the eastern states together.[38] Reports of this kind were frequently entered as evidence in debates on the adequacy of game laws.[39]

Most of this game passed from hunter to consumer through the well-organized market. St. Louis and Chicago were the primary collection points for western game from the south and north, respectively. Much of it was then forwarded to eastern markets, primarily to New York, which was the center of the substantial European trade. The New York market also received direct shipments from many regions on the eastern seaboard. Boston served as the distribution point for the trade in Maine venison. Other large markets thrived in Philadelphia, Baltimore, and Washington, D.C. Hunters brought their game to market in a variety of ways. Some were employed directly by dealers and worked on salary or commission.[40] Others regularly worked through the numerous commission agents who controlled the market in some regions. Still others shipped game directly to dealers in response to circulars.[41]

While evidence of the kind presented here is not conclusive, the general patterns in the markets for game can be summarized as follows: correcting for seasonal trends, demand and supply grew together during much of this period. As urban populations, which were without direct access to wildlife, increased, demand grew. At the same time, improvements in market-hunting technologies and in refrigerated storage and transport enabled markets to be supplied in greater abundance without increasing costs. As state restrictions on the exportation and sale of game began to be significantly enforced and as scarcity increased toward the end of the century, the supply of game to markets was reduced and prices began to rise. Within any season, prices were high as a species first came into market, owing to relative scarcity coupled with first-of-season demand. Prices fell as the season progressed and rose again as desirable species became scarce faster than demand was satiated.[42]

Some wildlife products and specialty items passed through

commercial channels outside of the game markets described here. Particularly significant were the trades in birds and their eggs and feathers. The major demand for birds and feathers came from the milliners who employed a wide variety of species in designing ornate arrangements on fashionable hats. One interested observer noted, in 1886, that 11 of 13 women passengers in a Madison Avenue horse car wore birds:

(1) heads and wings of three European starlings; (2) an entire bird (species unknown), of foreign origin; (3) seven warblers, representing four species; (4) a large tern; (5) the heads and wings of three shore-larks; (6) the wings of seven shore-larks, and grassfinches; (7) one-half of a gallinule; (8) a small tern; (9) a turtle-dove; (10) a vireo and a yellow-breasted chat; (11) ostrich-plumes.[43]

On another day, in the same year, the noted ornithologist Frank M. Chapman observed that of 700 women's hats spotted on the New York streets, 542 were decorated with 40 recognizable species of birds.[44]

A few hunters specialized in supplying plumage to the millinery trade, but they were not alone in incurring the wrath of the preservationists.[45] Ornithologist J. A. Allen, in summarizing the destruction of bird life in 1886, also indicted the "bad small boy" who shoots at birds for fun with pea shooter, sling-shot, or stone, and who robs nests; the foreign-born hunter who," to demonstrate to himself that he has really reached the 'land of the free,' equips himself with a cheap shotgun, some bird traps, clapnets, or drugged grain, one or all, and hies himself to the nearest haunt of birds for indiscriminate, often very quiet, slaughter or capture"; and the taxidermist who, though often engaged in quite legitimate work, is also often in business with milliners.[46]

For some species, the egg trade replaced the collection of the birds themselves as the major threat. Once a colony was located, existing eggs were destroyed to avoid taking those partially incubated or spoiled. Thereafter, all eggs over one inch in diameter were collected, except those of pelicans "whose eggs are too fishy for any stomach."[47] Particularly large trades were conducted from Laysan Island, of the Hawaiian group, and from the Farallones, thirty miles off of San Francisco. Gull and murre eggs from this latter area found a ready market in San Francisco, where they brought from twelve to twenty cents per dozen, slightly less than hens' eggs. Every second day during the two-month laying season beginning in mid-May, eggs were collected and shipped to the city. Between 1850 and 1856, 3 to 4 million eggs are reported to have been removed and sold. The harvest fell gradually to 92,000 in 1896, the final year of the trade.[48] A second drain on birds' eggs resulted from the demand for oölogical specimens. Children might earn considerable pocket money by filling orders for dealers who supplied collectors. A circular issued by one dealer in 1884 specified the following prices for sound eggs: robin, two cents; bluebird, three cents; song sparrow, two cents; baltimore oriole, five cents; meadow lark, six cents; peewee, eight cents, and rose-breasted grosbeak, twenty cents.[49]

WILDLIFE ON THE LAND

The distribution and abundance of wildlife on the land depends on complex ecological considerations including climate, vegetation, competition, and predation. Land use changes, primarily by altering vegetational patterns, require substantive adjustment by wildlife populations. Agriculture generally replaces a stable pattern of vegetation with a simplified pattern requiring considerable effort to maintain. Although its introduction is necessarily harmful to animal species requiring a climax community, other species will be favored. Clearing of wooded areas provides open spaces and, consequently, edge environments on the interface between clearing and the remaining natural vegetation. In these edges, the features of two habitat types are available to species of low mobility but with diverse vegetational requirements.[50] The quantity of edge environment produced is a function of the separation between cleared parcels and their shapes. Agriculture also provides supplementary food sources, both in the particular crops grown and in the pest species which may thrive on them. Finally, even in a region of complete clearing for agriculture, some species will find sanctuary in hedgerows, road margins, and other small but protected regions of natural vegetation.

One view of the rapid change in land use patterns is afforded by U.S. Census statistics (see Table 1) showing farm lands as a proportion of total acreage in several representative states and regions.[51] Of special note are the reforestation following the abandonment of farms in the North Atlantic states, and especially in northern New England, the extraordinarily rapid cultivation of the North Central states, and the vast majority of the far west remaining uncultivated by 1900.

Agriculture is not the only form of land transformation with consequences for wildlife, nor was this transformation limited by the federal government's technical ownership of the western territories. Huge tracts of public domain were mined, lumbered, and settled. The legacy of these practices includes private timber harvest in the national forests and the exercise of mining claims on federal lands, including those protected by the National Wilderness Preservation System.

Nor do these data include residential and industrial transformations, both of which disturb natural habitats to varying exents and which imply changes in the character of wildlife populations. To the extent that these activities reduce available land area or add to natural habitats elements which threaten life systems, the overall quantity of wildlife would be reduced. Its relative distribution may change to favor species that have a competitive advantage under resulting conditions. Thus, the English sparrow and the Norway rat thrive in urban environments.[52]

The success of the nascent preservationist movement partially offset the loss of habitat by development. The Wildlife Refuge System, founded with the designation of Pelican Island in 1903, is only the most pointed manifestation of a broad interest in protecting land from private development. The assembly of the national parks, begining with the creation of Yellowstone in 1872, and a network of national forest reserves beginning in 1891, equally illustrate the concern. Even the

Table 1 IMPROVED FARM LANDS AS A PROPORTION OF
TOTAL ACREAGE

	1850	1880	1900
United States	.058	.147	.214
North Atlantic	.313	.420	.360
Vermont	.423	.534	.346
Pennsylvania	.298	.463	.456
South Atlantic	.168	.203	.258
Virginia	.249	.326	.386
Florida	.009	.025	.040
North Central	.055	.279	.464
Ohio	.373	.685	.730
Illinois	.140	.723	.767
Minnesota	(a)	.135	.343
Iowa	.023	.298	.695
Kansas	(b)	.204	.476
South Central	.056	.125	.201
Tennessee	.191	.314	.379
Texas	.004	.074	.114
Western	(a)	.021	.036
Nevada	(b)	.005	.008
California	(a)	.105	.118
Colorado	(b)	.009	.034

SOURCE: *Twelfth Census of the United States, Agriculture, Part I* (Washington,
D.C.: U.S. Census Office, 1902), Table 52, pp. 692-93.

a) Acreage given but proportion is less than .001.
b) no acreage given.

initial transfer of the federal domain to state and private hands was reversed with the 1911 Weeks Act, which provided for the acquisition of national forest lands in the eastern states. The states, most notably New York, also began to acquire and manage vast tracts of undeveloped lands.

The designation of reserves guaranteed neither the safety of the wildlife within nor the maintenance of natural habitats. Yellowstone's big game went largely unprotected for more than twenty years following the creation of the park, and even after the passage of the 1894 Act to Protect the Birds and Animals in Yellowstone National Park, wildlife was exhibited for the entertainment of tourists. Predator control continues to be practiced on the public domain as the values associated with undisturbed wildlife populations are weighed against those of livestock, game animals, and even the recreational use of federal lands.

Whatever the observed consequences of these land use changes on wildlife, the ecological dynamics which produced them were little understood in the nineteenth century. Some of the earliest systematic data on the relationship between land use practices and wildlife populations was provided by Aldo Leopold in his *Report on A Game Survey of the North Central States*, published in 1931.[53] He considered game birds, and especially the quail, in greatest detail. He divided the habitat development of the quail into four stages: the presettlement stage, in which the bird is confined to the edges of prairies and open woods; the crude agricultural stage, with the introduction of grain crops, imported weeds, and fences of wood or hedge (these additional clearings extended the range of the species into the woods, and hedges extended it into the prairies); agricultural intensification, in which weedy, rail fences and hedges were replaced with barbwire, and woods were turned into pastures; and, finally, the stage of agricultural depression, good roads and automobiles, and the reversion of marginal lands. A sample farm in Calloway County, Missouri, was in the transition between stages two and three at the time of Leopold's survey. During a seven-year period, the 280 acre farm had its osage hedges removed,[54] fencerows debrushed, gullies cleared and filled, livestock increased, and woodlands pastured. The estimated number of quail fell from 210 to 90, or from 14 to 6 covies, if each contained 15 birds. The earlier density approached the one bird per acre which was the greatest that Leopold considered anywhere sustainable.[55]

In general, quail were scarce in the whole north central region before agriculture and had already begun to decline in Wisconsin by 1880 and in northern Illinois by 1887. The pinnated grouse (prairie chicken) exhibited a similar development. "In Illinois as late as 1836 a hunter was extremely lucky if he could bag a dozen [grouse] in a day. Some years later, with much less effort, one could have shot 50 in a day, and there were records of 100 to a single gun."[56] The species was already in decline there by the 1870's and had moved northward into the newer agricultural regions.

While much of the change in wildlife abundance was incidental to land transformation, other changes resulted more directly from the behavior of those who worked on the land. The individual landowner operated rationally with respect to the animals on his land only within the context of his knowledge about them and their relationship to his own enterprise. He may, for example, have killed birds which seemed to feed primarily on cultivated grain when in fact they fed primarily on weed seeds or insect pests. During the latter half of the nineteenth century, the scientific community gained a new sophistication in these matters. The natural environment began to be seen not as hostile and disordered, able to be subdued by brute force, but as a delicately balanced web of interconnections in which man must intrude gently, or at least with some forethought.[57]

Much of this new learning was made available to the farmer and landowner through the publications of the U. S. Department of Agriculture under the general rubric of "economic ornithology and mammology."[58] The first of these, published in 1856, even before agriculture was accorded departmental status, was Robert Kennicott's "The Quadrupeds of Illinois,

Injurious and Beneficial to the Farmer." In his introduction, Kennicott set forth a particularly perceptive view of the relationship between wildlife and the farmer.

However injurious wild animals may be to man, he should not forget that he himself is very often the cause of their undue destructiveness. When destroying his crops, they are only following the instincts with which they have been endowed by the Creator. Ruled by All-wise laws, every animal fills its appointed place exactly, existing not alone for itself, but forming a necessary part of the vast system of Nature. One class of animals keep in check certain plants; others prevent the too great increase of these, while those having few enemies are not prolific. Man interferes unwisely, and the order is broken. It is true that, to some extent, an interference with the natural regulations of the animal creation is necessary to the progress of civilization. Man appropriates to himself the food of many animals, but as they continue to devour this wherever found, he must therefore sometimes destroy them or lose his property.

But before waging war upon any animal, let us study its habits. . . . [59]

This study of habits generally brought forth for each species a catalog of behavioral characteristics which were assigned positive and negative values and were summed into a proclamation on the importance of the species to the landowner. Of the *Mustelidae* (skunk, badger, otter, weasel), for example, an 1863 report found:

. . . Birds and their eggs, and young, are often destroyed by them, but not to an extent at all balancing the benefit they do; nor is the injury caused by an occasional raid into the poultry yard at all comparable to the immense benefit we receive from their unceasing nightly labors in the destruction of rats, mice, and insects. [60]

Many of these papers dealt with birds and emphasized the eating habits of various species. Among the earliest studies of this kind was that conducted in Massachusetts on the robin. In 1857, the Massachusetts Horticultural Society petitioned the legislature for repeal of the law protecting the species on the grounds that it was a significant threat to cultivated fruits. A committee of three, led by Professor J. W. P. Jenks, was appointed to study the eating habits of the robin. It reported that although the species had some predilection for ripe cherries and strawberries, its diet, even at that time, was thoroughly mixed with a variety of insects. At other times, it was wholly composed of animal matter.[61] Protective legislation was retained. By the 1880s, analysis of stomach contents had become a well-known and popular means for establishing the respect owed various species of birds. By 1900, the Bureau of Biological Survey had collected some 32,000 bird stomachs and had analysed 14,000 of them, representing over 100 species.[62]

But it was the predators, birds and mammals, which excited the greatest controversy. Farmers and stockmen never questioned the belief that these creatures threatened the viability of their enterprise, nor did sportsmen seriously question the conventional wisdom that predators offered only competition in the hunt. Consequently, few opposed the bounty laws which encouraged the killing of hawks, eagles, owls, wolves, coyotes, and a variety of other species.[63]

In 1886, C. Hart Merriam, ornithologist and mammologist of the U.S. Department of Agriculture, publicly spoke out against the conventional wisdom on predators. He directed his remarks in particular at an 1885 Pennsylvania bounty of fifty cents (plus twenty cents to the official taking the affidavit) on each hawk, owl, weasel, and mink killed within the state. In the first year and a half, the state paid out $90,000 which Merriam calculated to represent 128,571 animals. In determining the real cost of the program, he assumed that only hawks and owls were involved, and that these birds would have killed 5,000 domestic fowl, worth twenty-five cents each, for an eighteen-month loss of $1,875. On the first level, then, the state spent forty-eight dollars to save one. But each predator would have killed 1,000 mice or their insect equivalent, and each of these would have caused two cents' damage per year. Thus, the removal of each bird permitted damage of thirty dollars over eighteen months, and the total kill of 128,571 represented a loss of $3,857,130, or $2,105 for every dollar saved. To this must be added the damage to be expected due to the growth of the rodent population following the removal of predators.[64] While Merriam's analysis shows considerable naiveté regarding the dynamics of wildlife populations, his point is generally well taken. Additional evidence on the food habits of these birds was submitted by B. H. Warren, the ornithologist of the Pennsylvania Department of Agriculture.[65] The "scalp act" was repealed, in no small measure due to the influence of these studies.

Several publications of the Department of Agriculture shed additional light on the problem of predatory birds. The most important of these, appearing in 1893, was A. K. Fisher's *The Hawks and Owls of the United States in Their Relation to Agriculture.*[66] It was clearly oriented toward the farmer's world view, stating that any species on balance harmful to his interests should not only be refused protection but should be actively destroyed. The study consisted of the consecutive considera-tion of several species, noting the general habits of each which aid and those which hinder the efforts of the farmer. Another of Fisher's essays, *Hawks and Owls from the Standpoint of the Farmer*, divided raptors into four categories: species wholly beneficial, species chiefly beneficial, species whose harmful and beneficial qualities balance, and species generally harmful. The majority of species were classified as chiefly beneficial.[67]

These studies can be faulted in minor ways, but they served to point out the discrepancies between the state of the law and the state of nature. Fisher attributed the difference largely to poor information:

It is to be regretted that the members of the legislative committees who draft State game laws are not better acquainted with the life histories of raptorial birds. . . . That the beneficial species of hawks and owls will eventually be protected there is not the slightest doubt, for when the farmer is convinced that they are his friends he will demand their protection.[68]

Education was slow in coming. In 1928, Alden H. Hadley surveyed opinions on the legal status of these birds in the forty-eight states. Of the thirty-nine state game commissioners responding to his questionnaire, only three felt that laws protecting certain hawks and owls were both desirable and practical; eleven were sympathetic but felt such laws were impractical owing to the lack of public support; and twenty-five opposed such legislation altogether. One eastern game official stated:

I would be opposed to a law giving protection to hawks and owls of any species. I consider these birds vermin in every sense of the word and I will always contend that the only good hawk is a dead hawk, which also applies to owls. I really do not believe that there is a magistrate in the state who would be in sympathy with a law to punish persons for killing hawks and owls even though it was a law, and I know there is not a court or jury in any of the counties of this state that would convict a person for killing a hawk or owl even though the law prohibited same on the statute books of this state.[69]

Some species were deliberately removed; others were deliberately introduced. Thomas Woodcock, president of the Natural History Society of Brooklyn, was reported to have brought over a number of European song birds in 1846, including goldfinches, linnets, bullfinches, and skylarks. Between 1872 and 1874, the Acclimatization Society of Cincinnati spent some $9,000 importing 1,200 birds of thirty-four varieties.[70] At about the same time, the Society for the Acclimatization of Foreign Birds liberated a similar selection in Cambridge, Massachusetts. In 1877, the American Acclimatization Society released a number of starlings in Central Park. This latter introduction brought praise from *Forest and Stream*, which is difficult to understand in light of the experience with the English sparrow.

The City of New York is greatly indebted to the American Acclimatization Society for the setting at liberty a large number of common starlings . . . in the Central Park. . . . From the moment of leaving the nest it begins to manifest its bright and joyous disposition by singing merrily all day, no matter how inclement the weather, nor how scanty its supply of food, teaching us a lesson of contentment more effectually than could some of our greatest philosophers.[71]

Though this attempt failed, another introduction, made in 1890, was successful. In 1889 and 1892, the Society for the Introduction of European Song Birds of Portland, Oregon, liberated birds of some twenty species. At the time of Palmer's survey in 1899, several species had apparently established themselves.[72] Only the starling and the English sparrow now survive in significant numbers.

More notable success was achieved with the importation of game birds. The important example during the period 1850 to 1900 is the ring-necked or Mongolian pheasant. Although a number of unsuccessful attempts to establish the species were made during colonial times, it was not until 1881, when Judge O. N. Denny, then consul-general at Shanghai, made a shipment of birds which was released in the Willamette Valley, that the species became well established in this country. In 1894, the Oregon State Board of Agriculture estimated that over a million of the species existed within the state.[73] The bird was easily established in other states of the west and north central regions, and soon became a popular game species. In South Dakota, perhaps the most successful habitat, it is estimated that between 1919, the year of the first open season, and 1940, 20 million of the birds were legally killed. The 1966 hunt was approximately 3 million.[74]

While the failure of the skylark to establish itself was a disappointment and the success of the ring-necked pheasant was welcomed, the success of the English sparrow was clearly a disaster. The bird, "like a noxious weed transplanted to a fertile soil [had] taken root and become disseminated over half a continent before the significance of its presence [had] come to be understood."[75] It was first introduced, though unsuccessfully, by Nicholas Pike and his associates at the Brooklyn Institute, who imported eight pairs in 1850. Other, more successful introductions followed, and the species spread rapidly. A very prolific breeder given the appropriate habitat, the sparrow was at home in urban America, where it found ample food in the undigested grain in horse droppings.[76]

The success of the importations was generally aided by the game laws, which failed to distinguish among song and insectivorous birds. Some knowledgeable individuals steadfastly supported the species well after the popular tide had begun to turn against it. John Galvin, city forester of Boston, claimed that the sparrows had nearly exterminated the "nasty yellow caterpillar" on the trees of the Common as well as the canker worm of the elms of the South End. At the time of his letter to *Forest and Stream*, he claimed that they were eating the small lice in elm buds. His men could do nothing in this way, for "they had no wings like the sparrow. . . . Thousands of dollars could not pay the city for their loss."[77] The shrike, or butcher bird as it was commonly called, was ordered destroyed in Boston, "thus thwarting, with characteristic human short-sightedness, the first efforts Nature made to readjust the disturbed balance of her forces."[78]

It was not the lack of preference for an insect diet which engendered general hatred of the bird but that it was considered vulgar and tended to displace other species, most notably the bluebird, with its own numbers. The generally held impression of the bird is reflected in Henry Van Dyke's characterization that the "kingdom of ornithology is divided into two departments—real birds and English sparrows."[79] Elliot Coues suggested in 1879 that an analysis of stomach contents be made in order to ascertain the true worth of the species. C. J. Maynard examined the stomachs of fifty-six sparrows taken from the central Boston area. Not one contained a trace of insect matter.[80]

One of the initial goals of the Department of Ornithology upon its creation in 1886 was a study of the English sparrow. In 1889, Walter B. Barrows, assistant to Merriam, compiled *The English Sparrow (Passer Domesticus) in North America, Especially in Its Relations to Agriculture*. The viewpoint of the study was made clear:

The magnitude of the evil and the absolute necessity of taking active and comprehensive measures for its abatement will be better understood after an examination of the following seven sections which precede the recommendations which we hope may lead finally to the extermination of the European House Sparrow in America.[81]

At this time, the recognition of the evil of the sparrow had barely registered in the game laws. Only two states provided incentive for their destruction with bounties, New York made their feeding or shelter a misdemeanor, and Massachusetts, Rhode Island, New Jersey, and Pennsylvania excluded them from the list of protected nongame birds.[82] Barrows's study recommended the repeal of all protective legislation, the killing of the birds and the destruction of their nests and eggs, and the arrest of anyone intentionally feeding them (except with a view toward their ultimate destruction by poison or trap), introducing them to a new location, or interfering with those in the act of destroying them.[83]

Although these recommendations were gradually adopted, they failed to achieve the extirpation of the species. Nonetheless, the struggle increased public awareness of a nature balanced so delicately that the introduction of a single species could disrupt the distribution and abundance of so many others; it provided a good test for the tools of economic ornithology and the analysis of stomach contents; and it was a cause around which groups such as the American Ornithological Union and the Audubon Society could organize and thereby gain support for other activities.

In the wake of the experience with the English sparrow, many observers grew concerned over its repetition with other species. The rapidity with which the rabbit gained pest status in Australia and the mongoose gained it in Jamaica were other examples to encourage the creation of controls on further importations into the United States. In 1886, Merriam suggested that the importation of all species other than domesticated animals, certain song birds, and those animals intended for exhibition in zoos and menageries, be placed under the control of the commissioner of agriculture.[84] Another plea from Merriam's bureau came in 1898 with the publication of "The Danger of Introducing Noxious Animals and Birds."[85] In 1897, *Forest and Stream* suggested the establishment of federal control under the Biological Survey.[86] In 1900, with the passage of the Lacey Act, the federal government acquired this general authority. The act specifically prohibited the importation of the mongoose, fruit bat, starling, and English sparrow, and gave to the secretary of agriculture the authority to exclude any other species in the interest of agriculture. Permits were required for importations outside of a small group of approved species. In the first five years following enactment, 1,563 permits allowed the importation of 2,841 mammals, 819,970 canaries, 30,837 game birds, and 154,928 miscellaneous birds. Permission to import was denied for 7 mongooses, 54 flying foxes, 2 starlings, 15 blue titmice, and 1 great titmouse.[87]

The impacts of sport hunting, market hunting, and land-use changes have been considered seriatim. For some species they offset one another and for others, they compounded change. This chapter closes with a brief summary of the ways in which these three activities impacted on the populations of the passenger pigeon and the bison. The rapid demise of these species forcefully illustrated the power of man in altering the abundance and distribution of wildlife populations. These changes could no longer be seen as merely fortuitous; it was evident that they followed logically from changes in land-use patterns and hunting intensity. It was likewise evident that deliberate policy might protect and restore wildlife populations.

The Passenger Pigeon

The passenger pigeon was primarily characterized by its dense occurrences, both in migration and in nesting. It was originally distributed over the northern portion of the United States, east of the Rockies, and into Canada. The flocks that astounded Wilson and Audubon were viewed in the forests of Indiana, Ohio, and Kentucky.[88] Within thirty years, the bird was uncommon in these regions, and major nestings occurred farther west, in Michigan, Wisconsin, Minnesota, and into the Dakotas. Thirty years more and major nestings were not seen again. They had moved, it was supposed, beyond mail and telegraphic communication, just as had the Indian and the buffalo. As unexplored regions rapidly diminished, it became clear that no such western migration had occurred. One alternative theory placed the demise of the huge flocks in a freak storm over the Caribbean. Others suggested their migration to

Australia or South America. By the 1890s, only small groups and isolated birds were sighted.

In spite of their obvious scarcity, extinction was not generally anticipated. In 1897, *Western Field and Stream* wrote that the birds were "as liable to return at any time as unexpectedly as they went."[89] *Forest and Stream*, in 1899, supposed that the species would "live long in the land, but never again as a bird found in enormous numbers."[90] Its reasoning was based on the theory that the fewer the birds, the less subject they were to danger from man. Apparently the natural occurrence of the species in extreme densities was viewed as merely fortuitous and not as essential to the well-being of the species. As late as 1906, John Burroughs claimed, on good evidence, the existence of a sizeable flock in Greene County, New York. "I have no doubt," he wrote, "that the wild pigeon is still with us, and that if protected we may yet see them in something like their numbers of 30 years ago."[91] The last member of the species, Martha, died in the Cincinnati Zoo on September 1, 1914, at the approximate age of twenty-nine years.

During their period of abundance, the huge flocks roosted and nested in deciduous forests in which food supplies were most abundant. They preferred beech nuts to other mast, but, when unable to choose, survived well on acorns, fruits and berries, and cultivated grains. The earliest New England settlers, and the Indians before them, found the passenger pigeon a most useful species. According to Chief Pokogan, "whole tribes would wigwam in the brooding places." Seldom killing the old, they were attracted to the squabs from which the squaws made "squab butter" by boiling off the large amounts of fat from the young birds. Other birds were smoked and dried for future use in much the same way as was buffalo meat.[92] Though they were hunted by New England settlers, significant commercial hunting did not begin until about 1840 and hunts of grand proportions until the 1860s.

The methods of harvest were selected primarily for their efficiency. The most popular was the large net which was sprung upon a mass of birds previously baited to the "bed."[93]

To lure birds from overhead flight, the "stool pigeon" was often employed. This was

> a live pigeon tied to a small circular framework of wood or wire attached to the end of a slender and elastic pole, which is raised or lowered by the trapper from his place of concealment by a stout cord and which causes constant fluttering. A good stool-pigeon (one which will stay upon the stool) is rather difficult to obtain, and is worth from $5 to $25. Many trappers use the same birds for several years.[94]

The harvesting of squabs competed strongly with the harvesting of mature birds. The young birds were knocked from nests with poles or picked off the ground after felling the trees in which the nests were lodged. H. B. Roney described the scene at the Petoskey nesting of 1878, where he found "a large force of Indians and boys at work, slashing down the timber and seizing the young birds as they fluttered down from the nest. As soon as caught, the heads were jerked off from the tender bodies with the hand, and the dead birds were tossed into heaps."[95] Local Indians shot broad-headed arrows at nests to knock out the young. When nestings occurred in birch forests, pigeon harvesters often set fire to the tree bark, which sent adult birds above the conflagration and the young to the ground in escape, whereupon they were collected. The squabs were so obese that many split open upon falling to the ground.

The overall magnitude of the pigeon harvest was a matter of some dispute. Isolated reports include the following: from Hartford, Michigan, in 1869, 3 carloads per day were shipped to market for forty days, which yielded a total of 11,880,000 pigeons. Another Michigan town was reported to have shipped some 15,840,000 birds over a two-year period.[96] From the Michigan nesting of 1874, a single railroad station is reported to have shipped 80 barrels per day, each containing from 30 to 50 dozen birds, for the length of the nesting season.[97] Two reports from the Shelby, Michigan, nesting of 1876 suggest that 350,000 and 398,000 birds were shipped per week.[98]

The last major nesting occurred near Petoskey, Michigan, in the spring and summer of 1878. Roney estimated a total shipment of 1,500,000 dead birds and 80,352 live from Petoskey station alone. To this he added an undetermined number shipped by water, express, and wagon, as well as those removed by hunters themselves. Finally, noting the "thousands of dead and wounded ones not secured, and the myriads of squabs dead in the nest by trapping off of the parent birds soon after hatching," he arrives at a remarkable minimum total of 1 billion birds "sacrificed to Mammon."[99] Though there was considerable dispute about these calculations, no one questioned that the Petoskey nesting led to a harvest of huge proportions.[100] Every pair of oxen available in the region was hired at $4.00 per load to haul birds into the town for shipment. "The road was carpeted with feathers, and the wings and feathers from the packing houses were used by the wagon load to fill up the mud holes in the road for miles out of town."[101]

The market for passenger pigeon was extremely well organized, especially during the Michigan nestings of the 1870s. Pigeoners were in telegraphic communication with one another and with eastern markets, so that information about roostings and market prices circulated quickly. The demand for dead birds derived wholly out of their use as food, those having been stall-fed bringing higher prices owing to tenderness and weight. Live birds were demanded primarily by sporting clubs for use in trap shooting. Market prices fluctuated greatly over the course of the nesting season and according to distance from the point of capture. In the 1870s, the price of pigeons fluctuated from $1.00 per dozen to $.12 per dozen at the nesting, depending on abundance. For Michigan birds which brought $.40-.50 in the field, Chicago prices were $.50-.60. Live birds bringing $.40-.60 in the field brought $1.00-2.00 in the city, owing to the care, food, and space required between capture and delivery.

The demise of the species followed the masive lumbering of the deciduous forests and the extensive hunt described here. Population density may have been a reproductive trigger so that the population was reduced, at least in its latter stages, more than in proportion to its direct reduction by hunting. The critical minimum population was reached before efforts aimed at preservation were effectively organized. [102]

The Bison ("Buffalo")

Whereas the passenger pigeon was not thought generally to be threatened by the massive harvests of the 1870s and 1880s, most observers expected, and many looked forward to, the decimation of the American bison population. Most generally, the rapid decline of the species resulted from the profitability of its capture at a time when its range was demanded for settlement (see Map 1). While the Indian could coexist with the buffalo (and had for many centuries), the white man could coexist with neither the bison nor the Indian. Allowing the bison herds to be destroyed undermined the nomadic life of the Plains Indians and facilitated their confinement to reservations. [103]

The pre-white impact of the Indians on the bison population was minimal. While massive slaughters were occasionally undertaken by driving herds into pens or over precipices, much of the meat was preserved as jerky or pemmican, and the hides, bones, and cartilage were judiciously employed as weapons and cultural artifacts. [104] Even these large slaughters, perhaps on the order of 1,000, were not as destructive to the species as were natural catastrophes. Thousands of animals were drowned as herds crossed rivers coated with thin ice; thousands more became mired in mud and sand. [105]

The dependence of whites on bison began with the fur trappers of the Old Northwest who, having learned from the Indians, subsisted on buffalo meat during forays into the wilderness. [106] From the early nineteenth century, the demand for buffalo robes was substantial. The softened hides, with full winter hair, were considered by some to be necessary for survival on cold winter sleigh and wagon rides. Smaller robes were made into overcoats and old bull robes into overshoes with the wool on the inside.

Although "sportsmen" publicly ridiculed the idea of hunting buffalo, many found it difficult to pass up the opportunity

Map 1. RANGE OF THE AMERICAN BISON

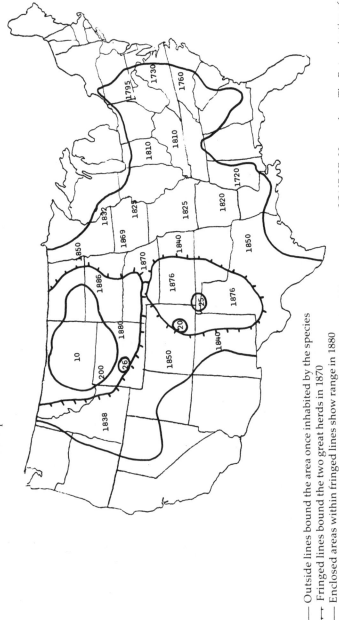

SOURCE: Hornaday, *The Extermination of the American Bison*.

Outside lines bound the area once inhabited by the species
Fringed lines bound the two great herds in 1870
Enclosed areas within fringed lines show range in 1880
Numbers represent remaining wild animals as of January 1, 1889
Dates indicate last sighting in region

to chase after the beast and add it to the list of species met and taken. Access for the hunter improved as the railroads pushed into the plains. Special excursions were organized:

RAILWAY EXCURSION
AND
BUFFALO HUNT

An excursion train will leave Leavenworth, at 8 A.M., and Lawrence at 10 A.M., for

Sheridan,

on Tuesday, October 27, 1868, and return on Friday. This train will stop at the principal stations both going and coming. Ample time will be had for a grand Buffalo

Hunt on the Plains

Buffaloes are so numerous along the road that they are shot from the cars nearly every day. On our last excursion our party killed twenty buffaloes in a hunt of six hours.

All passengers can have refreshments on the cars at reasonable prices. Tickets of round trip from Leavenworth, $10.00. [107]

The construction of the railroads brought another demand for the buffalo—as food for the work crews. Hunters were salaried by the companies to provide meat for the camps. William Frederick Cody (Buffalo Bill) was a major purveyor for the Kansas Pacific's 1,200 workers. [108] In his eighteen months of work for the railroad, he claims to have killed some 4,862 buffalo, an average of just under nine per day. [109]

Were this pattern of hunting for subsistence and for sport to have continued, the buffalo's fate would have been slower in coming but certain nonetheless. The construction of the transcontinental railroad itself divided the population into a northern and a southern herd, never again to unite. Stringing the prairies with barbed wire would, by cutting off migration routes and water sources, have greatly reduced the effective range of the species. But this sequence of events was not allowed to play itself out. First, tanning technology caught up with and then surpassed that of the Indians; not only could suitable robes be produced, but poor quality robes and summer hides could be tanned for use as leather. This made the hunt profitable year round. Second, transportation costs were reduced as the railroads moved farther and farther into buffalo territory. Third, with improved weapons, the species could be shot cheaply and from a distance, and so the hunter could pick the herd off one by one without causing a stampede. Finally, refrigerated cars and smoke houses made even the meat marketable.

Primarily owing to ease of access, the southern herd was the first to be harvested.[110] Throughout western Kansas, little occurred during the years 1871-74 that was not related to the trade in buffalo. Martin S. Garretson estimated that two-thirds of the 4,000 residents of Dodge City in 1873 were buffalo hunters.[111] Entry into the industry was easy, and, for the landless, turning buffalo hunter was the quickest way to earn a living. In 1870, bull hides brought the hunter $2.00 each, and cow and calf hides $1.75 each. Tongues, that portion of the meat most generally marketed, brought $.25 on the range and $.50 in eastern markets.[112] The "mop," or hair which falls over the animals' horns and eyes, brought $.75 per pound.[113] Finally, the bones of animals previously killed for hides and tongues were profitably collected. In 1875 it was reported that heads and ribs, to be ground into fertilizer, brought $5.00 per ton; shins and shoulder blades, convertible into carbon for sugar refining, brought $10.00 per ton; and horns, the tips of which were sawed off for use in the umbrella, fan, and pipe trades, brought $30.00 per ton. A Kansas City journal noted that "the trade has been opportune for the settlers, as it brought them the means of a livelihood when the crops failed."[114] Lincoln Ellsworth claims that between 1868 and 1881, some $2,500,000 was paid for bones. At an average price of $8.00 per ton and with 100 skeletons to the ton, the remains of 31,250,000 bison were required.[115]

The hunt was organized by merchants along the route of travel between the hunting grounds and the marketing and processing towns of Kansas. These entrepreneurs, either by advancing equipment in exchange for a percentage of the hides taken or by organizing hunting teams on salary, came to control the trade. The usual hunting party employed by a merchant contained four or five men. A five-member party included one shooter, two skinners, a cook, and a stretcher.

The best overall statistics on the destruction of the southern herd are provided by Dodge. He was successful, however, in gaining accurate statistics of shipment only from the Atchison, Topeka, and Santa Fe Railroad. Other companies declined on the grounds that clerical time could not be spared. Dodge suspected that it was a fear of bad publicity which motivated the silence. His figures are based on the assumption that the other railroads, in particular the Kansas Pacific and Union Pacific, carried in sum twice the cargo of the Santa Fe. They show that during the years 1872, 1873, and 1874, 1,378,359 skins, 6,751,200 pounds of meat, and 32,380,359 pounds of bone were shipped from the southern herd. To the estimated number of skins, Dodge added an estimate of the number of buffalo killed for market but whose skins were spoiled or not taken, the number killed by Indians, by sportsmen, and by other hunters shipping through Hudson's Bay Company or to California. He arrived at a three-year total of nearly 5,500,000 buffalo killed.[116]

The northern herd, which had been harvested by Indians through the 1870s, was finally decimated by white hunters between 1882 and 1884 following the western extension of the Northern Pacific Railway. The course of the trade is difficult to document owing to the lack of separate railroad statistics for buffalo hides and parts. Fort Benton was a major center of the trade, and as early as the mid-1870s, T. C. Powers and I. G. Baker, the two largest dealers in the region, were sending their own hunting parties into the herds three times a week. The shipment from the region between 1874 and 1877 ranged between 80,000 and 100,000 per year.[117] Hornaday estimated the size of the northern herd in 1870 at 4,000,000 below the Platte River and 1,500,000 above.[118] By 1883, there were almost none.

Curiously enough, not even the buffalo-hunters themselves were at the time aware of the fact that the end of the hunting season of 1882-'83 was also the end of the buffalo. . . . In the autumn of 1883 they nearly outfitted as usual, often at an expense of many hundreds of dollars, and blithely sought "the range" that had up to that time been so prolific in robes. The end was in nearly every case the same—total failure and bankruptcy.[119]

By 1886, robes were so scarce in the region that they were imported from New York to Ft. Benton bringing a price of twenty-five to thirty dollars.[120]

In that same year, Hornaday, representing the Smithsonian Museum, led an expedition in search of bison for the museum's natural history collection. The hunt, in the region between the Yellowstone and Missouri rivers, provided 25 new specimens, not an insubstantial portion of the wild individuals then to be found in the United States.[121] By 1889, the remaining wild buffalo numbered 85.[122] In addition, some 256 were held in captivity, and another 200 were "protected" in Yellowstone National Park. Combined with an estimated 550 animals in the Northwest Territory of Canada, the total living bison numbered 1,091.[123] By the turn of the century, only about 20 animals, part of the Yellowstone herd, remained in the United States in any state resembling the wild.

NOTES

1. Remarks by W. Y. W. Ripley of the association, quoted in Vermont, *Report of the Commissioners of Fish and Game, 1893/94*, pp. 58-59.

2. Ibid., p. 57.

3. Vermont, *Report of the Commissioners of Fish and Game, 1897/98*, p. 68. Exaggerated reports even reached Paris, where the *Revue Scientifique* wrote that Vermont deer were so numerous and tame that "it is scarcely practicable to drive them away" [John W. Titcomb, "Fish and Game in Vermont," in Edward A. Samuels, ed., *With Rod and Gun in New England and the Maritime Provinces* (Boston: Samuels and Kimball, 1897), p. 333].

4. Ibid., pp. 66-70. Recent evidence suggests illegal kills are large. A Wisconsin study indicated that in a county in which 1,714 legal bucks were taken, 3,000 illegal does and fawns were left in the woods. A New York study concluded that of every 101 deer taken in the state, 34 were taken legally, 14 were killed by cars and dogs, and 53

were taken illegally, both in and out of season [Paul Tillett, *Doe Day: The Antlerless Deer Controversy in New Jersey* (New Brunswick: Rutgers University Press, 1963), p. 23].

5. Nathaniel R. Dickenson and Lawrence E. Garland, *The White-Tailed Deer Resources of Vermont* (Montpelier: The Vermont Fish and Game Department, Agency of Environmental Conservation, 1974), p. 36.

6. Vermont Fish and Game Department, *Vermont's 1975 Game Annual* (Montpelier, 1975), pp. 9-12. With the introduction of an antlerless season in 1979, harvests again rose. An estimated 24,000 deer were taken in Vermont during the 1980 season [*Brattleboro Reformer* (December 8, 1980), p. 1. See Chapter 6, note 181].

7. Guidebooks available to the sportsman-tourist were uniform in the caution they issued concerning Indian attacks. Charles H. Sweetser, in his *Tourists' and Invalids' Guide to the Northwest*, suggests that sportsmen unsatisfied with the opportunities in Minnesota wait until "sundry Pacific railroads have made matters comparatively safe" in the Dakotas before indulging in the "luxury of free-and-easy hunting" there. For one who can't wait, he notes that "the handful of white men in Dacota [sic], and the whole population of Minnesota, will not blame you a bit, if you crack your rifle not far from the scalp of one of these Sioux butchers who slaughtered men, women, and children, indiscriminately, at Farmers' Village and New Ulm. The chances are, however, that Sir Red-Skin will get a first pop at your precious cranium" (New York: The Evening Mail, 1868, p. 37).

8. Great Northern Railway, *Shooting and Fishing along the Line of The Great Northern Railway*, 6th ed. (St. Paul, 1903), p. 6. Other examples of railroad publications for the sportsman include the Boston and Maine's *Directory of Guides in Fish and Game Country* (1909); Pere Marquette's *Fishing and Hunting in Michigan* (Detroit, 1902); the Southern Railway's *Hunting and Fishing in the South* (Washington, D.C., 1904); Southern Pacific's *California Game "Marked Down"* (San Francisco, 1896); Chicago, Milwaukee, and St. Paul's *Fishing and Hunting Along the Line of the Chicago, Milwaukee and St. Paul Railway* (1899); the Missouri Pacific Railway's *Ideal Hunting and Fishing Grounds* (1900); and the Bangor and Aroostook Railroad's *Big Game and Fishing Guide to North-Eastern Maine* (Bangor, Maine: Printed by R. A. Supply Co., Boston, 1898). The railroads were not alone in providing guides for the sportsman. See, for example, the United States Cartridge Company's *Where to Hunt American Game* (Concord, N. H.: United States Cartridge Co., Rumford Press, 1898); and the J. Stevens Arms and Tool Company's *Guns and Gunning* (Chicopee Falls, Mass., 1908).

9. Although Hallock regards "concealment a virtue no longer," he refrained from entering certain haunts of woodcock, snipe, trout, and salmon. "These shall be held as sacred from intrusion as the penetralia of the Vestals" (*Sportsman's Gazateer*, Preface). Other guides were *Fur, Fin, and Feather: A Compilation of the Game Laws of the Different States and Provinces of the United States and Canada; to which is added a List of Hunting and Fishing Localities, and other Useful Information for Gunners and Anglers*, revised and corrected for 1871-72 (New York: M. B. Brown & Co., 1871), and William C. Harris, *The Sportsman's Guide to the Hunting and Shooting Grounds of the United States and Canada* (New York: The Anglers' Publishing Company, Charles T. Dillingham, 1888).

10. James Colles, *Journal of a Hunting Excursion to Louis Lake, 1851* (Blue Mountain Lake, New York: Adirondack Museum, 1961). For a comprehensive historical survey, see Frank Graham, Jr., *The Adirondack Park: A Political History* (New York: Alfred A. Knopf, 1978).

11. John B. Bachelder, *Popular Resorts and How to Reach Them*, 2d ed. (Boston: John B. Bachelder, 1874), p. 144.

12. *Forest and Stream* 2 (June 18, 1874), 290. Even for Murray, the "region is now ruined for the lover of solitude and nature.... A man cannot now sit down on the border of Tupper's Lake to fish a quiet hour, without becoming the center of a score of interested and fashionably dressed spectators" (quoted in Graham, *The Adirondack Park*, pp. 65-66).

13. This system also has the virtue of reducing rural unemployment. On the constitutionality of this and other legislation that discriminates against nonresidents, see Chapter 5, note 62.

14. *Forest and Stream* 51 (August 6, 1898), 107.

15. *Forest and Stream* 20 (May 3, 1883), 261.

16. *Forest and Stream* 3 (November 26, 1874), 249. This view was sustained by Adirondack guide Alvah Dunning. "... I druther they'd stay out o' my woods. They'll come anyhow, an' I might as well guide 'em fer if I don't some un else will, but I druther they'd keep their money and stay out of the woods. I can make a livin' without 'em, an' they'd starve to death here without me" [Fred Mather, *My Angling Friends* (New York: Forest and Stream Publishing Co., 1901), pp. 125-26].

17. Thomas Alexander, *Game Birds*, Vol. 28, No. 571 of the Seaside Library (New York: George Munro, 1879), p. 19.

18. John Bumstead, *On the Wing: A Book for Sportsmen* (Boston: Fields, Osgood, and Co., 1869), p. 78.

19. E. J. Lewis, *Hints to Sportsmen* (Philadelphia: Lea and Blanchard, 1851), p. 94.

20. Richard Irving Dodge, *The Plains of the Great West and Their Inhabitants, Being a Description of the Plains, Game, and Indians, & c. of the Great North American Desert* (New York: G. P. Putnam's Sons, 1877), p. 104.

21. *Forest and Stream* 36 (January 29, 1891), 21. David E. Lantz noted that by 1905, community coyote drives had become a "popular feature of rural sport in some parts of the country." U. S. Department of Agriculture, Division of Economic Ornithology and Mammology, *Coyotes and Their Economic Relations*, Bulletin No. 20 (Washington, D.C.: G.P.O., 1905), p. 21.

22. *Forest and Stream* 50 (April 23, 1898), 321.

23. G. O. Shields, ed., *The Big Game of North America* (Chicago: Rand McNally & Co., 1890), p. 546. The market hunting of this species was under way as early as 1877 with the organization of the Florida Swamp Crocodile Hide Tanning and Oil Refinery Company, Ltd., which took 9,000 of the animals in Louisiana worth seventy-five cents per hide, plus the value of the oil [*Forest and Stream* 10 (April 11, 1878), 182]. Prices were little changed in the 1890s when marketable skins, three to twelve feet long, brought an average of sixty cents. Small, live specimens, sold to tourists, primarily in Jacksonville, brought ten to thirty dollars a hundred. *Forest and Stream* estimated that the decade of the 1880's saw 2.5 million alligators killed in Florida alone [Ibid., 40 (June 8, 1893), 489]. The 1890 census estimated that the hunt yielded, in that year, $169,593 from the sale of 5,645 eggs, 2,219 pounds of ivory, 1,097 gallons of oil, 19,925 live animals, and 168,122 skins [*Report on Statistics of Fisheries in the United States* (Washington, D. C., 1894), p. 37].

24. Robert B. Roosevelt, *The Game Birds of the Coasts and Lakes of the Northern States of America* (Boston: Carleton, 1866), p. 65.

25. *Forest and Stream* 21 (August 2, 1883), 1.

26. The large bag, another mark of the unscrupulous market hunter, was not so readily dismissed. As late as 1890, *Forest and Stream* commented, " . . . We may feel a certain kindly regard for the man who shoots a few birds and then stops for fear of ruining the chances of later arrivals; but we should regard the act rather as a virtue of supererrogation than as of ethical obligation" [35 (October 9, 1890), 225].

27. Letter to *Forest and Stream* 36 (December 10, 1891), 411.

28. George Shiras, III, *Hunting Wild Life with Camera and Flashlight*,

2d ed., 2 vols. (Washington, D. C.: The National Geographic Society, 1936), pp. 57, 60. Originally published in *National Geographic Magazine*, July 1906.

29. L. W. Brownell, *Photography for the Sportsman Naturalist* (New York: Macmillan, 1904), p. 21.

30. Thomas F. DeVoe, *The Market Assistant, Containing a Brief Description of Every Article of Human Food Sold in the Public Markets of the Cities of New York, Boston, Philadelphia, and Brooklyn* (New York: Orange Judd Publishing Company, 1866).

31. Peter Mattheissen, *Wildlife in America* (New York: The Viking Press, 1959), p. 166. Another menu may be found in *Forest and Stream* 9 (December 6, 1877), 352.

32. *American Sportsman* 3 (December 27, 1874), 204.

33. John E. Skinner, *An Historical Review of the Fish and Wildlife Resources of the San Francisco Bay Area*, Water Projects Branch Report #1, The Resources Agency of the California Department of Fish and Game (Sacramento, June 1962), Appendix G-2, p. 210.

34. *Forest and Stream* 3 (November 19, 1874), 230. See Tober, "Allocation of Wildlife Resources," Tables 4.1-4.3, for additional information on retail game prices.

35. *Forest and Stream* 3 (December 17, 1874), 294.

36. *American Sportsman* 1 (March 1872), 7; *Forest and Stream* 13 (December 25, 1879), 934. See Tober, "Allocation of Wildlife Resources," Table 4.4, for additional information on fur and skin prices.

37. *Forest and Stream* 10 (May 23, 1878), 309; *American Sportsman* 2 (August 1873), 172. See Tober, "Allocation of Wildlife Resources," Table 4.5, for a compilation of selected individual hunts.

38. *Forest and Stream* 16 (March 31, 1881), 169; U.S.D.A. Bureau of Biological Survey, *Progress in Game Protection in 1910*, by T. S. Palmer and Henry Oldys, Circular No. 80 (Washington, D. C.: G.P.O., 1910), pp. 9-10.

39. See, for example, the discussion in Chapter 6 below on the relationship between hounding deer and the size of the deer population.

40. Chicago dealers, for example, sent teams of five or six into the field after prairie chickens (*The New York Times*, October 13, 1882, p. 4). See also David and Jim Kimball, *The Market Hunter* (Minneapolis: Dillon Press, 1969).

41. Not every market hunter was able to find a willing buyer. Webb recounted the reaction received from the Fulton Market dealers in New York upon his shipment of buffalo meat from the plains:

In the first quarter, it carried dyspepsias and disgust, and was so tough that the recipients, with the utmost effort, could not find a tender regret for our danger in obtaining it; while our New York consignee wrote that the first morning's steaks "finished the market," and very nearly finished his customers. He found it impossible, even by the Fulton Market method of subtraction, to get three hundred dollars worth of express charges out of half that amount of sales, and suggested a discontinuance of shipments [W. E. Webb, *Buffalo Land: An Authentic Account of the Discoveries, Adventures and Mishaps of a Scientific and Sporting Party in the Wild West* (Cincinnati: E. Hannaford and Co., 1872), p. 249].

42. Some examples of market glut occur in the literature. In 1874, venison and "bar" meat were overflowing the game stands in Memphis, and, in Shreveport, citizens were "living on luscious mallards at from ten to fifteen cents each" [*American Sportsman* 5 (December 12, 1874), 168]. In 1876, the Texas market was so well supplied that the price of quail had fallen to fifty cents per dozen and "the boy who sells them will remove their heads for nothing, and say 'thank you' besides" [Ibid., 9 (December 16, 1876), 169].

43. *Science* 7 (February 26, 1886), 197.

44. *Forest and Stream* 26 (February 25, 1886), 84.

45. *The Report on Statistics of Fisheries in the United States* from the census of 1890 includes data on the extent of bird hunting in the Gulf Coast states. These figures, the source of which is not made clear, show that 345,731 birds worth a total of $86,707 were captured in Texas, Louisiana, and Florida in the year 1889. The largest number of a single named species was 29,453 white crane. Other species mentioned were sandhill crane, bittern, egret, flamingo, heron, ibis, and osprey. (Three-fourths of the birds were classed as "other.") Of the 1,516 persons employed in the trade, 1,108 were classed as white, 18 as colored, 115 as Indian, 17 as foreign born, and 258 of unknown characteristics [U.S. Census Office, Department of the Interior, *Report on Statistics of Fisheries in the United States at the Eleventh Census: 1890* (Washington, D. C.: G.P.O., 1894), p. 37]. See also U.S. Department of Agriculture *Yearbook, 1899* "Review of Economic Ornithology in the United States," by T. S. Palmer (Washington, D.C.: G.P.O., 1899).

46. J. A. Allen, "The Present Wholesale Destruction of Bird-Life in the United States," *Science* 7 (February 26, 1886), 193. See also Robin W.Doughty, *Feather Fashions and Bird Preservation: A Study in Nature Protection* (Berkeley: University of California Press, 1975).

47. George B. Sennett, "Destruction of the Eggs of Birds for Food," *Science* 7 (February 26, 1886), 199-201.

48. The Farallone Egg Company controlled the nesting site until 1881, when it was dispossessed by the California Light House Board whose employees subsequently managed the trade for their own profit. In 1897, under public pressure, the board stopped the trade. U.S. Department of Agriculture *Yearbook, 1899*, "Review of Economic Ornithology in the United States," by T. S. Palmer, pp. 271-72; *Forest and Stream*, 47 (October 3, 1896), 261; 48 (March 6, 1897), 181; see also the articles by Walter E. Bryant and by Leverett M. Loomis in the *Proceedings of the California Academy of Sciences*, 2d ser., Vol. 1 (1888), pp. 31-36, and Vol. 6 (1896), pp. 356-58.

49. *Forest and Stream* 22 (April 10, 1884), 203.

50. See Raymond Dassman, *Wildlife Biology* (New York: John Wiley & Sons, 1964), pp. 74-75.

51. The proportions are computed from statistics on improved land areas from *Twelfth Census of the United States, Agriculture, Part I* (Washington, D.C.: U.S. Census Office, 1902), Table 52, pp. 692-93.

52. In some man-made environments, not even these species survive, as this 1887 report indicates.

The Loyalhanna River, of Pennsylvania, is just now in a suitable condition of filth and corruption to point a moral. There is less water in it than at any time within several years, and all the nastiness which the dwellers along its banks see fit to dump and drain into it just stays there to putrify and poison water and air. Vitriol from paper mills, spent liquor from tannery vats, sewage from towns, and sundry other ingredients compose a liquid medium in which the fish indigenous to the originally pure waters of the Loyalhanna do not thrive. Suckers, perch, and catfish, together with the bass which were put into the stream by the State Fish Commission, have perished, and are piled up along the shores, screens and tail-races for miles, where they threaten a pestilence on a large scale. The inhabitants are said not to relish the evil which has come upon them in this day of reckoning; but they who dance must pay the piper, they who dig pits for themselves must fall therein, and they who convert a stream of pure water into a sewer and transform a blessing of bounteous nature into a conduit of filth must expect some time to have their nostrils filled with the stench thereof. And the beauty of it all is that in spite of this lesson the Loyalhanna folks will go right on draining their tanneries and paper mills and sewers into the stream and making all ready for another pestilence, whenever the clouds of heaven again refuse to purify the river. That is human nature, the world over [*Forest and Stream* 28 (June 21, 1887), 549].

53. The *Survey* (Madison, Wisconsin, 1931) was made for the Sporting Arms and Ammunition Manufacturers' Institute under the direction of its Committee on Restoration and Protection of Game. In his preface, Leopold commented that its motive "hardly requires explanation: success in game restoration means continuance of the industry; failure in game restoration means its shrinkage and ultimate liquidation" (ibid., p. 5). On Leopold's life and work, see Susan L. Flader, *Thinking Like a Mountain* (Columbia: University of Missouri Press, 1974).

54. Osage hedges were planted extensively beginning in the 1860s. They were favored, especially on the prairie where wood was scarce, because they were essentially free, requiring only labor costs, and because, when grown, they provided impenetrable and durable boundaries. As the root systems spread, reducing the yield of crops, and with the discovery that the hedge served as the host of the San Jose scale, an insect pest of orchards, the hedges were removed. Power tractors facilitated the process. "Within a single decade [1910-1920] the osage hedge virtually disappeared" (Leopold, *Report on a Game Survey*, pp. 64-65).

55. Ibid., pp. 24-42.

56. U.S. Department of Agriculture. Bureau of Biological Survey, *The Grouse and Wild Turkeys of the United States, and Their Economic Value*, by Sylvester D. Judd, Bulletin No. 24 (Washington, D.C.: G.P.O., 1905), p. 12.

57. This was the theme of George Perkins Marsh's *Man and Nature*, published in 1864. In a letter to Spencer Fullerton Baird in 1860, Marsh called the book in progress "a little volume showing that whereas [others] think that the earth made man, man in fact made the earth" [quoted by David Lowenthal in his introduction to a 1965 edition of the work (Cambridge, Mass.: Belknap Press of Harvard University Press), p. ix].

58. These papers appeared in two groups. The first group was published in the late 1850s and early 1860s in the *Reports of the Commissioner of Agriculture* (*Reports of the Commissioner of Patents*, before 1862). It included Ezekial Holmes, "Birds Injurious to Agriculture" (1856), E. Michener, "Insectivorous Birds of Chester County, Pennsyl-

vania" (1863), E. A. Samuels, "Mammology and Ornithology of New England, with Reference to Agricultural Economy" (1863), Samuels, "Oology of Some of the Land Birds of New England as a Means of Identifying Injurious or Beneficial Species," (1864), Samuels, "The Value of Birds on the Farm" (1867), J. R. Dodge, "Birds and Bird Laws" (1864), and D. G. Elliot, "The 'Game Birds' of the United States" (1864). The second group, following the official entry of the Department of Agriculture into matters of ornithology and mammology in 1886, appeared in several places. Papers issued before 1900 include eleven Division of Biological Survey bulletins, sixteen papers in the series entitled North American Fauna, twenty-two Circulars of the Division of Biological Survey, twelve essays in the *Yearbook* of the Department of Agriculture, plus the material included in the *Report of the Commissioner of Agriculture*. Several titles indicate the range of subject matter of these publications: Walter Barrows, *The English Sparrow* (Bulletin 1, 1889), A. K. Fisher, *The Hawks and Owls of the United States in Their Relation to Agriculture* (Bulletin 3, 1893), Barrows, *The Common Crow of the United States* (Bulletin 6, 1895), C. Hart Merriam, *Brief Directions for the Measurement of Small Mammals and the Preparation of Musuem Skins* (Circular 11, 1889), Merriam, *Directions for Collecting the Stomachs of Birds* (Circular 12, 1891), Merriam, "The Geographic Distribution of Animals and Plants in North America" (*Yearbook*, 1896), and Palmer, "The Danger of Introducing Noxious Animals and Birds" (*Yearbook*, 1898). A complete list of these publications prior to 1928 may be found in Jenks Cameron, *The Bureau of Biological Survey: Its History, Activities and Organization*, Institute for Government Research of the Brookings Institution. Service Monographs of the United States Government, No. 54 (Baltimore: Johns Hopkins University Press, 1929), pp. 217-51.

59. U.S. Patent Office, *Report of the Commissioner of Patents for the Year 1856: Agriculture*, pp. 52-53. Parts 2 and 3 of Kennicott's study may be found in the *Reports* for the years 1857 (pp. 72-107), and 1858 (pp. 241-56).

60. U.S. Department of Agriculture, *Report of the Commissioner for the Year 1863*, p. 268. A major difficulty with this type of analysis is that it is at the same time possible for a species to be of net benefit to society, to the agricultural community, to a town, or to a neighborhood in the sense that the combination of losses it prevents and gains it provides are greater than the direct losses it causes, while it is of net harm to a particular individual or to another group. To attempt to persuade the chicken rancher to refrain from killing skunks and weasels because, even though they cause him substantial loss, the damage they prevent

in his neighbors' wheat fields is large, is to ask him to behave irrationally from the point of view of managing his own enterprise. Without compensation for the losses he must bear, there is no reason to expect such behavior on a voluntary basis.

61. Solon Robinson, *Facts for Farmers*, 2 vols. (New York: A. J. Johnson, 1866), I, 184-90; U.S. Department of Agriculture, *Report of the Commissioner for the Year 1863*, "Mammology and Ornithology of New England," by E. A. Samuels, p. 282.

62. The Department of Ornithology under the Division of Entomology of the Department of Agriculture was created in 1885 largely through the efforts of the American Ornithological Union. The AOU was formed in 1883 as a national outgrowth of the Nuttall Ornithological Club, based in Cambridge, Massachusetts, and organized in 1873. The Union formed three major research committees—on faunal areas, on the problem of the English sparrow, and on bird migration. Unable to pursue these matters with its limited budget, the Union sought governmental aid. Through the efforts of Spencer Baird of the Smithsonian Institution and the U.S. Commission of Fish and Fisheries, a hearing was gained before the House Agricultural Committee. The eventual result was the appropriation of $5,000 for the "investigation of the food habits, distribution, and migrations of North American birds and mammals in relation to agriculture, horticulture, and forestry . . . " to the aforementioned department. (The Division of Entomology had, in turn, been formed in 1877, following the outbreak of locusts in the west earlier in the decade. It was originally attached to the Geological and Geographical Survey as the U.S. Entomological Commission, but was transferred to the Department of Agriculture in 1881.) C. Hart Merriam, on the recommendation of the AOU, was appointed to head the agency, a position he retained for twenty-five years. In 1886, the department gained division status as the Division of Economic Ornithology and Mammology; in 1896, its scope was officially broadened as the Division of Biological Survey; and in 1905, it achieved Bureau status. In 1940, the bureau was merged with the Bureau of Fisheries to become the Fish and Wildlife Service under the Department of the Interior. See Cameron, *The Bureau of Biological Survey*, passim. The 1956 Fish and Wildlife Act divided the Fish and Wildlife Service into the Bureau of Commercial Fisheries and the Bureau of Sport Fisheries and Wildlife. In 1970, the former bureau was transferred to the Department of Commerce and, in 1974, the latter became the U.S. Fish and Wildlife Service.

63. Nor were bounties for agricultural pests uncommon. Montana,

for example, paid five-cent bounties on 698,971 squirrels in the winter of 1886, and Minnesota collected 280,000 quarts of grasshoppers at ten cents each during a single week in 1875 [*Forest and Stream* 29 (September 22, 1887), 161; *Forest and Stream* 4 (June 24, 1875), 314. For a compilation of selected bounty laws, see James Tober, "The Allocation of Wildlife Resources," Table 4.9, pp. 214-15].

64. U.S. Department of Agriculture, *Report of the Commissioner of Agriculture for the Year 1886*, "Report of the Ornithologist and Mammologist," pp. 228-29.

65. The results of Warren's study were published in U.S. Department of Agriculture, *Report of the Commissioner of Agriculture for the Year 1887*, "Report of the Ornithologist and Mammologist," pp. 402-22.

66. U.S. Department of Agriculture, Division of Ornithology and Mammology, Biological Survey Bulletin Number 3.

67. U.S. Department of Agriculture, Bureau of Biological Survey, Biological Survey Circular Number 61 (July 18, 1907). See also U.S. Department of Agriculture, Bureau of Biological Survey, *The North American Eagles and Their Economic Relations*, by Harry C. Oberholser. Bulletin Number 27 (Washington, D.C.: G.P.O., 1906).

68. U.S. Department of Agriculture, Biological Survey Circular Number 61, p. 2.

69. Alden H. Hadley, "The Legal Status of Hawks and Owls: A Statistical Study," *Transactions of the Fifteenth National Game Conference*, 1928, p. 46.

70. "This society grew out of the yearnings of a few cultivated Germans to hear once more, in their own groves and gardens, the familiar bird songs of the Fatherland" (*The New York Times*, May 8, 1874, p. 9).

71. *Forest and Stream* 8 (June 14, 1877), 307.

72. U.S. Department of Agriculture, *Yearbook of the Department of Agriculture, 1899*, "Review of Economic Ornithology in the United States," by T. S. Palmer, pp. 287-90.

73. *Forest and Stream* 42 (January 6, 1894), 1.

74. George Laycock, *The Alien Animals* (Garden City, N.Y.: Published for the American Museum of Natural History by the Natural History Press, 1966), pp. 35-39.

75. U.S. Department of Agriculture, Division of Economic Ornithology and Mammology, Biological Survey Bulletin Number 1, p. 22.

76. Barrows claimed that up to six nestings per year with five or six eggs per set were possible (U.S. Department of Agriculture, Biological Survey Bulletin Number 1, pp. 27-29). It should be noted that the climax

population of sparrows declined with improved street cleaning and with the shift away from draft animals and toward motorized transportation.

77. *Forest and Stream*, 8 (May 31, 1877), 261. In the same issue there appeared several letters expressing views that ranged from mild to vehement disagreement.

78. Letter of Elliot Coues in *Field and Forest*, reprinted in *Forest and Stream* 7 (July 12, 1877), 329. The shrike was one of the few known enemies of the English sparrow.

79. *Fisherman's Luck* (New York: Scribner's, 1899), pp. 57-58. See also Peter J. Schmitt, "Call of the Wild: The Arcadian Myth in Urban America, 1900-1930" (Ph.D. diss., University of Minnesota, 1966), pp. 80-81.

80. *Forest and Stream* 12 (April 10, 1879), 190.

81. U.S. Department of Agriculture, Biological Survey Bulletin Number 1, p. 40.

82. The Ohio bounty provided a payment of ten cents per dozen and that of Michigan, a penny each (ibid., pp. 150, 167-73).

83. Ibid., p. 150.

84. U.S. Department of Agriculture, *Report of 1886*, p. 258.

85. U.S. Department of Agriculture, *Yearbook of 1898*, pp. 87-110.

86. *Forest and Stream* 48 (April 3, 1897), 261.

87. U.S. Department of Agriculture, *Yearbook: 1905*, "Federal Game Protection: A Five Years' Retrospect," by T. S. Palmer (Washington, D.C.: G.P.O., 1906), pp. 544-45. See Chapter 6, text at notes 188-200, on the passage of the Lacey Act. The act, through several amendments, remains among the more important federal wildlife statutes.

88. Wilson described one flock he viewed "in passing between Frankfort and the Indiana Territory":

> If we suppose this column to have been one mile in breadth, (and I believe it to have been much more,) and that it moved at the rate of one mile in a minute, four hours, the time it continued passing, would make its whole length two hundred and forty miles. Again, supposing that each square yard of this moving body comprehended three Pigeons, the square yards in the whole space, multiplied by three, would give two thousand two hundred and thirty millions, two hundred and seventy-two thousand Pigeons!—an almost inconceivable multitude, and

yet probably far below the actual amount. [Alexander Wilson, *American Ornithology*, ed. T. M. Brewer (Boston: Otis, Broaders, and Company, 1840), p. 399].

89. *Western Field and Stream* 2 (June 1897), 46.

90. *Forest and Stream* 52 (February 4, 1899), 81.

91. Quoted from *Forest and Stream* in W. B. Mershon, ed., *The Passenger Pigeon* (New York: The Outing Publishing Company, 1907), p. 180.

92. Simon Pokagon, "The Wild Pigeon of North America," in Mershon, ed., *The Passenger Pigeon*, p. 54, originally printed in *The Chautauquan*, Vol. 20 (November 1895).

93. Nets six by thirty feet might have caught forty to fifty dozen birds per strike. See H. B. Roney, "Efforts to Check the Slaughter," Henry T. Phillips, "Notes of a Vanished Industry," and William Brewster, "Netting the Pigeons," all in Mershon, ed., *The Passenger Pigeon*.

94. Roney, "Efforts to Check the Slaughter," in Mershon, ed., *The Passenger Pigeon*, p. 80.

95. Ibid., p. 87.

96. C. Gordon Hewitt, *The Conservation of the Wildlife of Canada* (New York: Charles Scribner's Sons, 1921), p. 21.

97. *American Sportsman* 4 (September 5, 1874), 362.

98. *Forest and Stream* 8 (March 18, 1877), 69; *American Sportsman* 8 (June 3, 1876), 148. All of these numbers are large, but it is likely that a healthy population, with adequate habitat for nesting and feeding, could have sustained a harvest of this size. The mourning dove, which has become an important game species in a number of states, sustained a legal hunt in 1960 of 30 million, up from 15 million in 1949 and 19 million in 1955-56 (Harold S. Peters, "The Past Status and Management of the Morning Dove," *North American Wildlife Conference*, 1961, pp. 371-74).

99. Roney, "Efforts to Check the Slaughter," in Mershon, ed., *The Passenger Pigeon*, p. 92.

100. One critic placed the figure at a remarkably precise 1,107,866. He suggested that the overestimate resulted from an unfamiliarity with large numbers [E. T. Martin, "The Pigeon Butcher's Defense,' in Mershon, *The Passenger Pigeon*, pp. 99-100, reprinted from *American Field* (January 25, 1879)].

101. Roney, "Efforts to Check the Slaughter," in Mershon, ed., *The Passenger Pigeon*, p. 90.

102. See Vinzenz Ziswiler, *Extinct and Vanishing Animals*, rev. English ed. by Fred and Pille Bunnell (New York: Springer-Verlag, 1967), pp. 57-58, and A. W. Schorger, *The Passenger Pigeon: Its Natural History and Extinction* (Madison: University of Wisconsin Press, 1955).

103. The case was explicitly made in Congressional debate. In consideration of an 1876 bill to protect the bison in the territories from recreational hunting, Representative Throckmorton of Texas argued that the legislation "would work a great hardship on large portions of our frontier people who are in the habit of hunting buffaloes not only for food but for amusement." And he went on: "Now, sir, there is no question that, so long as there are millions of buffaloes in the West, so long the Indians cannot be controlled, even by the strong arm of the Government. I believe it would be a great step forward in the civilization of the Indians and the preservation of peace on the border if there was not a buffalo in existence" (*Congressional Record*, February 23, 1876, p. 1239).

104. William Hornaday, in his comprehensive study of the bison, placed substantial blame on the Indian for the extermination of the species.

> I have yet to learn of an instance wherein an Indian refrained from excessive slaughter of game through motives of economy, or care for the future, or prejudice against wastefulness. From all accounts, the quantity of game killed by an Indian has always been limited by two conditions only—the lack of energy to kill more, or lack of more game to be killed. White men delight in the chase, and kill for the "sport" it yields, regardless of the effort involved. Indeed, to a genuine sportsman, nothing in hunting is "sport" which is not obtained at the cost of great labor. An Indian does not view the matter in that light, and when he has killed enough to supply his wants, he stops, because he sees no reason why he should exert himself any further. This has given rise to the statement, so often repeated, that the Indian killed only enough buffaloes to supply his wants [*The Extermination of the American Bison, with A Sketch of Its Discovery and Life History*, The Annual Report of the U.S. National Museum (Washington, D.C., 1887), pp. 506-7].

By 1913, he had revised his views and considered large kills by Indians as prudent owing to the uncertainties of nature (*Our Vanishing Wildlife*, p. 8).

105. See Tom McHugh, "Bison Travels, Bison Travails," *Audubon* 74 (November 1972), 22-31, for a description of natural hazards befalling the species.

106. See R. O. Merriman, "The Bison and the Fur Trade," *Bulletin of the Departments of History and Political and Economic Science in Queen's University*, No. 53 (Kingston, Ontario, 1926), for a description of the dependence of western Candian trappers on the buffalo.

107. Excursion of The Kansas Pacific Railroad, quoted in E. Douglas Branch, *The Hunting of the Buffalo* (New York: D. Appleton and Company, 1929), p. 129.

108. Billy Comstock, chief scout at Fort Wallace, challenged Cody to a buffalo hunt over rights to the name Buffalo Bill. In three individual contests, taking place twenty miles east of Sheridan, Cody killed a total of sixty-nine to Comstock's forty-six [William F. Cody, *True Tales of the Plains* (New York: Cupples & Leon Company, 1908), pp. 74-78].

109. Cody used a wagon with four mules, a driver, and two butchers in his work. He lost five assistants (three drivers and two butchers) to the Indians (pp. 64-68).

110. By the Treaty of Medicine Lodge, signed in 1867, white hunters were prohibited from crossing below the Arkansas River. This territory was set aside for the Cheyenees, Arapahoes, Kiowas, and Comanches as a hunting ground "so long as the buffalo may range thereon in such numbers as to justify the chase." While the treaty was respected until about 1870, it was clear before that time that neither peace nor boundaries would last. See Dee Brown, *Bury My Heart at Wounded Knee: An Indian History of the American West* (New York: Bantam Books, 1971), Chapter 11.

111. Martin S. Garretson, *The American Bison* (New York: New York Zoological Society, 1938), p.114.

112. Other portions were, however, marketed.

The meat market opens in November, when the weather becomes cool enough for its transportation, and continues until the first of April. During these five months as much as 9,000,000 pounds are shipped from the Kansas prairies to all parts of the country. In the winter months, a buffalo-steak can be obtained about as easily and almost as cheaply in the butcher's stalls of the leading Northern cities, as a beef-steak or a mutton-chop, and in Kansas it is common as hog-meat. When Buffaloes are killed for this meat, only the hams and shoulders are brought in, and shipments are usually made in that shape, the hide nearly always being left on to the end of the journey. The leading markets for buffalo meat "in the rough" are St. Louis, Chicago, and Indianapolis, whence it is reshipped in cleaner and more artistic condition, to cities of the seaboard. At Kansas City, too, large quantities are cured and packed for Eastern use, and some successful experiments have been made in shipping direct to New York and Philadelphia from extreme Western Kansas in refrigerator cars. The price in Kansas ranges from $50 to $80 per ton in bulk; and the local dealers retail it at six to eight cents per pound [*American Sportsman* 5 (December 19, 1874), 182].

113. The Buffalo Wool Company was established in the Red River Settlement in 1821 with the aid of a British subsidy. It was found, however, that what cost two pounds, ten shillings per yard to make would bring only four shillings, seven pence, in England (Hornaday, *American Bison*, p. 450).

114. Quoted in *American Sportsman* 6 (September 11, 1875), 363.

115. Lincoln Ellsworth, *The Last Wild Buffalo Hunt* (New York, privately printed, 1916), pp. 7-8.

116. Dodge, *The Plains of the Great West*, pp. 140-43.

117. Branch, *The Hunting of the Buffalo*, p. 204.

118. Hornaday, *American Bison*, p. 504.

119. Ibid., p. 512.

120. *Forest and Stream* 25 (January 21, 1886), 501.

121. *Forest and Stream* claimed a total of thirty-three for Hornaday's hunt [28 (March 3, 1887), 106].

122. See Map 1. Four buffalo, located in western Dakota, are not shown.

123. Hornaday, *American Bison*, p. 525.

4

PROPERTY RIGHTS
AND MARKETS

In America the owners of game and the men who shoot it
belong (so very generally that the exceptions need not be
regarded) to two distinct and almost antagonistic sections
of the community—the plain farmer on the one hand, and
the better class of townsfolk on the other.

from an Englishman's
survey of American game
preservation, 1888

Whatever the solution to the increasing scarcity of wildlife, it
inevitably would involve some conflict between the "plain
farmer" and the "better class of townsfolk." The customary
liberties of the landowner and of the hunter could not but
threaten each other as the number of hunters grew, as agriculture spread and intensified, and as the object of the hunt
became more valuable. Sportsmen sought to reduce the conflict
by distinguishing themselves from other classes of hunters
whom they accused of abusing private property, but any gains
registered were wiped out by their support for the game-law
movement, which was viewed by rural populations as class
legislation for the benefit of the urban elite.

Everything hinged on the ownership of wild animals. If
wildlife were recognized as common property, could common
owners then pursue animals wherever they sought refuge,
even on private property? How much control could the state,
as trustee for common owners, exercise over the pursuit?
Alternatively, if ownership in wild animals were recognized as
private property, what kinds of voluntary exchanges would
emerge and how would they be enforced? In fact, there was

neither customary nor legal precedent for private ownership, the states had assumed ownership of wildlife, and the courts sustained that view. Nonetheless, the merits and demerits of private property in wild animals received considerable attention in the nineteenth-century sporting press and elsewhere.

Private property rights may be linked either to species or to land area and, if to land area, may be held either by the landowner or by another proprietor. These property designations similarly provide for the exchange of rights, through market transactions or otherwise, although the exchanges would be accompanied by very different transactional and policing costs. When the number of owners is large and when the populations over which they have jurisdiction intermingle, property rights in particular animals and their offspring can be maintained only at significant cost. The history of the livestock industry in the western United States is especially revealing in this regard.[1] There is some precedent even for private property of wild animals in ownership of "royal species" such as the swan and the whale by the English king. English law also offers precedent for the separation of property rights in land use from property rights in wildlife on the land. The costs which this system of royal grants and prerogatives imposed on tenant farmers was the subject of heated nineteenth-century debate.[2] Although the merits of these structures of property rights were occasionally argued in the United States, only the assignment of ownership in wildlife to the user of the land was paid significant attention.

But there are difficulties with this approach. The fugitive nature of wildlife resources suggests that, for any landowner to capture the gains from rational management, his land area must correspond closely to the habitat of identifiable populations of relevant species. While the owner of ten acres of woodland may be able to manage a population of small rodents, he could not do so with deer. Indeed, the owner of such a parcel might endeavor to capture whatever deer wandered through his property. In a community of small landholders, this is but a version of common property. The owner of a larger parcel of land might fence in a number of deer hoping to establish a viable population that could be regularly harvested. Were deer useful only as producers of meat and hides, this property arrangement might bring about management practices with the best interests of society in mind. But deer are also sport hunting targets, symbols of recreation quality, and indicators of ecosystem health, and these demands could not be efficiently met except on the largest parcels. With limited exceptions, American patterns of private land ownership have never been conducive to the efficient management of wildlife by landowners.

Historically, the more important constraint is distributional. Granting ownership of wildlife to the landowner provides him with a substantial windfall gain, and it gives him control over those who wish access. The debate over the assignment of rights in wild animals focused variously on both efficiency and equity considerations. "O.H.H." of Iowa wrote to *American Sportsman* that any laws designed to give landowners special claim to wildlife on their lands would only deny access to those who could not afford to pay for it in the field or at the market stall. "Game is one of the many gifts of nature to mankind in general, therefore, any law which compels a man to *buy* his game either before or after he shoots it is an unjust one. . . . [T]he fact that the game was *bought* would greatly spoil the pleasure of shooting even if the money were no object."[3] This was echoed by "A Vermonter" who argued that any efforts to preserve game at the expense of creating social divisions based on wealth should be ruled out. "Better *every* game bird and fish be destroyed" than to allow landowners, "who appear to think they ought to be the only ones who should have the right to hunt and fish," to post their lands. The new legislation in Vermont prohibiting trespass could not but facilitate the transfer of prime hunting grounds into the hands of a few wealthy citizens who would bar access to the rightful owners of the game.[4]

Inevitably others opposed all controls over the landowner. Pioneer fish culturist Fred Mather argued "this is my land, and my empire, as my house is my castle." Owning the land gave him the right "to cut down any tree or plant one as I see fit, to control the lives of all animals upon it." No one should have privileges to "come and kill anything upon it without my permission nor to dictate to me what I should kill, nor when."[5] Others were more magnanimous, pointing to the "evils of common property."

> If it was the common law that hens, chickens, turkeys, and geese belonged to nobody; that anybody might kill and eat them at pleasure, it would not be long before the race of domestic fowls would become extinct. . . . Or if it was the common law that sheep and lambs belonged to nobody, it would be impossible to preserve them from utter destruction. Each man, when he saw a sheep or lamb, would take and sequester it for his own use, lest his neighbor should get the start on him.
>
> There is no common or statute law protecting fish and game, therefore our fish and game are rapidly disappearing. . . .
>
> What we need is a law, not simply protecting game and fish . . . but one making game and fish the property of the owners of land on which they are found, and the streams through whose territory they run. . . .[6]

More conciliatory positions recognized the need for compromise. "It is indeed, right and proper that the farmer or landowner should be amply protected and guaranteed the full enjoyment of the fruits of all his labors. No law should be passed taking from him even the least of his actual rights, neither should any be enacted conferring upon him the rights of others."[7] This, of course, only begs the question of with whom the rights lay.

The historical value of free access precluded the assignment of property in wildlife to the landowner or to any other individual on equity grounds, and the increasingly strong, though not well defined, view of wise use and conservation, precluded it on efficiency grounds. At the same time, landowners were suffering damage from the activities of hunters on private lands. As the number of hunters increased, the damage resulting from trampled crops, the escape of livestock, and the destruction of improvements grew. As late as 1871, only five states had enacted legislation which in some way protected the landowner from unauthorized trespass for the purpose of hunting. (Seven others discriminated in favor of landowners by prohibiting trespass for the purpose of violating the game laws.) By 1900, most states and territories prohibited unauthorized trespass at all times. The form of the law differed. In some states, only fenced property was protected. In others, land wth growing crops was protected and, in still others, appropriately placed signboards were sufficient to guarantee privacy.

These provisions, by assigning control over access to wildlife resources to private parties, provided the same basis for private or market solutions to the problem of wildlife scarcity as did assignment of private ownership in the resource itself. In numerous cases, however, the first response of the landowner was to prohibit all access to hunters. Thus Plymouth, Connecticut, landowners posted their lands in 1875: "Notice, to whom it may concern forbidding all persons hunting, trapping or killing any partridge, quail, wood-cocks or any squirrels, rabbits or any species of game on our lands after this date." *American Sportsman* viewed this action as an unfortunate misunderstanding that would stop only gentlemen sportsmen, who obey the law, but not poachers and pothunters, who are responsible for all damage to private lands.[8] Similarly, eighty-one farmers in Winnebago County, Wisconsin, together posted their lands against hunters in reaction to extensive depredations. The group offered a $25.00 reward for information leading to the conviction of trespassers.[9] Some land was posted to the amazement rather than to the dismay of sportsmen. In Virginia and North Carolina, one correspondent found "old barren meadows, where a field lark would starve; swamps that a prowling coon would turn up his nose at, piny woods that nothing that runs on four legs, or birds of any feather could exist, all posted."[10]

In the south, the welcome accorded visiting sportsmen depended on particulars. Northern hunters were treated well if they possessed letters of introduction from prominent local citizens. Otherwise, a stranger was easily spotted and quickly forced out.

About the third day he will be waited upon by a delegation of local sportsmen, who will politely inform him that they have not preserved the game on their plantations for unknown persons to kill and ship away for money, and that he will do well to leave and stand not on the order of his going, but to go at once. This warning is usually all that is necessary. If another is needed, it will be of a nature and character that its recipient will remember.[11]

A considerable amount of land was posted against hunting in the south by the turn of the century. The Southern Railroad's *Hunting and Fishing in the South* (1904) showed that over half of 402 towns in eight states were largely posted.

The outright prohibition of access to wildlife by landowners not only prevented the damage which accompanied trespass but preserved for them the benefits of the wildlife which was useful in the fields and on the table.[12] Although the increasing appreciation of the value of wildlife on the farm made the farmer all the more protective of his property, it also induced him to consider its value to others and indeed to provide access to wildlife for those hunters willing to pay accordingly. T. S. Palmer advised that access might be "rented by the day or by the season, and may be accompanied with charges for board and lodging, the use of a team, or the time of a boy to act as guide, and will then net a very profitable return."[13] Thus, an association of sportsmen in Berks County, Pennsylvania, leased nearly 3,000 acres from fifteen farmers for $5.00 per year per farmer and received in exchange exclusive hunting and stocking privileges and half of all penalties collected for unauthorized trespass.[14] Similarly, the Farmer's Mutual Protective Union organized in New Jersey to keep track of local hunters. Each of the cooperating landowners issued hunting permits which were valid on the lands of all cooperators. The issuer was responsible for all damage caused by his "guests," and he kept a black list of those considered to be bad risks. The system opened some 15,000 acres to hunting. While there was no prescribed fee for a permit, nothing prohibited the individual landowner from making arrangements involving payment.[15]

These practices were not widespread. The distance between landowners and urban sportsmen was simply too great. *Forest and Stream*, in an effort to match landowners with sportsmen, offered its columns as a clearing house in 1880. While some response was elicited from sportsmen, nothing was heard from landowners.[16] In 1888, A. G. Bradley, the author of the chapter epigraph, observed, in his survey of game laws in the United States, that

so far no financial solution to the question has, I believe, been thought of; things are hardly ripe for that yet. Fond of money, and much in need of it as the American farmer generally is, the idea of taking it for the right of shooting would for many reasons require a good deal of digestion before it became a thoroughly accepted one.[17]

Some part of the early failure of this idea must be attributed to property damage which hunters forced upon landowners. Sportsmen issued voluminous propaganda aimed at distinguishing themselves from market hunters and other uncivilized types.

The genuine and honorable sportsman is the friend and ally of the agriculturalist. He will be found always ready to protect birds which are useful, to destroy the rapacious and hurtful, to prevent trespass, and enforce the laws. He pursues his favorite game at a season when the harvest is gathered in and the fields can be traveled without injury, and he does it in a manner that no reasonable man can complain of. If he does otherwise, he is no true "brother of the gun."[18]

Similarly, "Mohawk" distinguished between "sportsmen" who always ask permission to hunt, and "shooters" who not only fail to ask permission, but often damage property.[19] The market hunters, on the other hand,

are a peculiar class of hoodlums, made up in great part of men without an occupation, and among them we find the skedaddler, smuggler, thief, firebug, and lazy squatter who lives from what lumber he can steal, berries he can pick, fires he can fight, *after setting them*, or anything save honest labor. The middleman is the prototype of the city pawnbroker and junk dealer. . . . . [20]

By the turn of the century, southern European immigrants had joined the market hunter as the scapegoat of the sportsman.

These newcomers, ignorant both of the language and of the law, frequently mistake liberty for license, and, free from the restraints to which they were accustomed in their own country, imagine that they can hunt birds of all kinds without restriction as to place or season. When thus engaged, they throw down rail fences, trample grain, steal fruit, or commit similar depredations, and meet remonstrance or interference on the part of the owner with stolid indifference or a resentment which occasionally leads to personal encounter.[21]

This campaign was simply not very successful. The difficulty which sportsmen encountered in gaining access to desired hunting ground by lease increased the popularity of the private shooting ground.[22] The sportsman-owned and controlled preserve was held up not only as a means to exclusive access to game but also as a means to rationalize resource use by providing incentives for responsible management.[23] A few even viewed it as a generally desirable institution, modeled on the European system. Horace Greeley hoped, in his *Recollections of a Busy Life*, that

our children will see, though I shall not, the greater portion of Pike and Monroe Counties, with other sterile mountain districts of eastern Pennsylvania, converted into spacious deerparks of fifty to five hundred square miles each, enclosed by massive stone walls, and intersected by belts of grass traversing each tiny valley (so as speedily to stop the running of any fires that might chance to be started), planted with the best timber, and held by large companies of shareholders for sporting, under proper regulations.[24]

Greeley might have seen the beginning of his dream in the establishment of the Blooming Grove Park Association, chartered under Pennsylvania law in 1871. Located in Pike County and founded largely by the efforts of Charles Hallock and Fayette S. Giles, the park, at its inception, included some 30,000 acres. Part of the land was divided into separate regions for the protection of breeding populations of game species. The first such area, of 720 acres, was set aside for the white-tailed deer. Others were planned for elk, moose, and antelope, contracts for the collection of which were let to hunters in the north and west. The park was to include hotel accommodations and space for the members to build their own homes.[25] Hallock saw as its primary objectives the

> importing, acclimating, propagating, and preserving of all game animals, fur-bearing animals, birds, and fishes adapted to the climate; the affording of facilities for hunting, shooting, fishing and boating to members on their own grounds; the establishment of minkeries, otteries, aviaries, etc.; the supplying of spawn of fish, young fish, game animals, or birds, to other Associations or to individuals; the cultivation of forests; and the selling of timber and surplus game of all kinds.[26]

Considerable uncertainty surrounds the later history of the association. In 1873, William Ferguson, a member, wrote to the *Spirit of the Times* complaining of fiscal mismanagement and the absence of improvements. Those few which had been made were "built for show, and in cheap and nasty style. . . . One visit to the ground will convince any true sportsman that at present the shooting is a sham and the fishing a farce."[27] Other references to the club indicate its precarious financial condition. The $450 initial membership fee had fallen to $50, and by 1880 the club's indebtedness had risen to $71,000. But the association survived this period, however, and *Club Men of New York: 1901-02* lists Blooming Grove as one of a small number of associations to which important New Yorkers belonged.[28]

The Adirondacks, because of its good supply of game and its proximity to eastern population centers, was among the most intensively used sporting regions in the country, and a number of preserves were maintained there. The largest, controlled by the Adirondack League Club, reportedly contained 200,000 acres, 104,000 of which were owned outright.[29] *Forest and Stream* calculated that, in 1894, thirty-two clubs controlled 824,112 acres, all within the limits of Adirondack State Park.[30] Another report in that same year claimed that 565,000 acres of Adirondack lands were posted and fenced as private parks.[31] A 1902 study of the region by the Bureau of Biological Survey concluded that 791,208 acres were held within sixty private preserves, and an additional 1,153,414 acres were held by the state in the regional park.[32]

Other sporting preserves were located in prime waterfowl areas. Principal coastal preserves were found at the south shore of Long Island, the upper Chesapeake Bay, tidewater Virginia, the Currituck Sound,[33] and the Suisan marshes of California; inland preserves were maintained along the Illinois, lower Mississippi, Sacramento, and San Juaquin rivers and at Michigan's St. Clair Flats.[34] *Forest and Stream* saw no future for waterfowl in the absence of control of the prime grounds by well-managed clubs. A particularly hopeful sign was the Kitty Hawk Bay Club which controlled fifty miles of North Carolina shorefront and which "will, by reason of its vested rights, become a leading factor in obtaining such legislation as is now needed for the preservation and protection of game."[35]

These market solutions, based on the historical sanctity of private property, provided some realignment of access to wildlife based on a diversity of demands, but not enough. They also created new problems that undermined some of the gains. Just as monopoly in business enterprise by wealthy entrepreneurs was increasingly viewed as a limitation on the rights of the common man to enjoy the fruits of individual enterprise in a rich commercial environment, so monopoly in prime hunting grounds was viewed as a limitation on the rights of the common man to enjoy the fruits of individual enterprise in a rich natural environment. "Didymus" denied that

a few men, simply because fortune has given them the means, have the *moral* right to buy up thousands of acres and monopolize the sport to the exclusion of hundreds of others who are equally fond of it, but have not the means to buy the privilege. . . .

As wealth increases every acre of good shooting ground will be cut off from "the people," and where will be our boasted freedom?

One of our great financiers—an ex-Secretary of the Treasury—emphatically predicts that, in the not far future, this country will see a terrible revolution, the result of the tendency to concentration and monopoly; and he may be right, for we profess to be a free people, and will not tamely submit to oppression.[36]

Thirty years later, Henry Chase reported that

we hear many worthy Americans advocating the establishment of these preserves in this country. Have they wholly forgotten the lesson of history? Do they desire to foist the European system of game protection upon us? Or is it their wish to stir up all the class hatred they can in this country?[37]

But the preserve failed even to meet the needs of an influential group of sportsmen. Wilderness hunters were appalled by the tameness that had come to characterize British sport and feared the same outcome in the United States. Theodore Roosevelt, articulating widely held views, wrote that "shooting in a private game preserve is but a dismal parody; the manliest and healthiest features of the sport are lost with the change of conditions." His solution, " in the interest of the community at large," is a "rigid system of game laws rigidly enforced," and "great national forest reserves, which shall also be breeding grounds and nurseries for wild game."[38] The answer did not lie in increased reliance on the voluntary exchange of the market place but in significant government limitations on the freedom of individual hunters and in a reversal of a century-long public policy of land disposal.

The idea of public preserves for the maintenance of game in its natural habitat can be traced at least to George Catlin who, upon contemplating the fate of the bison and the Indian,

imagines them as they *might* in future be seen, (by some great protecting policy of government) preserved in their pristine beauty and wildness, in a *magnificent park*, where the world could see for ages to come, the native Indian in his classic attire, galloping his wild horse, with sinewy bow, and shield and lance, amid the fleeting herds of elk and buffaloes. What a beautiful and thrilling specimen for America to preserve and hold up to the view of her refined citizens and the world, in future ages! A *nation's Park*, containing man and beast, in all the wild and freshness of their nature's beauty![39]

Catlin's vision was hardly realized in the extensive network of Indian reservations which, by 1880, covered some 150 million acres of western lands. More hopeful was the reserva-

tion of national parks and forests, slowing the wholesale conversion of the public domain into farms and homesteads, and
offering some potential for the preservation of wildlife habitat,
if not of the wildlife itself.[40] Specific protection for wildlife on
public lands came first with the 1894 Act to Protect the Birds and
Animals in Yellowstone National Park, the principles of which
were then extended to the expanding park system, and especially with Roosevelt's executive order of 1903 creating the
Pelican Island bird reservation which would become but the
first sanctuary in the National Wildlife Refuge System.[41]

While the protection of wildlife on federal lands is largely a
twentieth-century story, the game law has a venerable history
in the United States stretching back to the early colonial
period. The interest in this control mechanism fell off considerably in early statehood, indicating perhaps that the game
laws had never significantly constrained behavior, that they
had arrived in the New World merely as a part of Old World
baggage, and, in some cases, that protected wildlife was sufficiently scarce as to make statutory protection redundant.
Owing largely to the efforts of sportsmen, with the support of
the scientific community and the newly organized fish and
game commissions, the game law attracted considerable support in the second half of the nineteenth century, and, by 1900,
most states had a constitutionally enforceable, though often
unenforced, set of statutes that nominally protected many
species of wildlife.

The statutes seemed to be directed to the market hunter, but,
politically more significant, they would, if enforced, nearly
circumscribe the customary freedom of landowners and tenants. The hostility of the landowner was founded in the
supposedly erroneous belief "that the sportsman desires to
prevent [wildlife] from being killed, even by the owners of the
soil, in order that, at certain seasons of the year, the country
may be invaded by himself and friends sallying from the towns
to gratify their own pleasure in the destruction of game, to the
attempted exclusion of everyone else."[42] Similarly, Solon
Robinson, writing for the agricultural community, hoped that
farmers would correct their mistaken view that game legislation was but a "device cunningly framed for depriving them of

their own natural and indefeasible rights.... In a word, they looked upon the Game Laws as an offensive, aristocratic, unrepublican, European invention; a sort of scheme for making the rich richer, and the poor poorer...."[43]

There is no doubt that sportsmen stood to gain by passing legislation restricting the hunting behavior of landowners and rural populations, but they would gain with a clear conscience. *Forest and Stream* editorialized in "We the People" that game laws favor no particular class but only protect all classes from certain lawless elements in society. Those favoring game legislation were "business and professional men, tradesmen, mechanics, property owners and farmers—the respectable portion of the community—*the people*...On the one side is respectability, thrift and worth; on the other, lawlessness, shiftlessness, vagabondage. There is no disputing the correctness of this classification."[44]

By the turn of the century, the urban press and the increasingly professional state fish and game commissions had joined sportsmen in their support for the notion that the game laws not only were of general benefit but were indeed of primary benefit to the lower classes. The drive for limited access to wildlife represented nothing less than the responsible, even self-sacrificing exercise of political power in husbanding the nation's remaining wealth in wildlife. The frontispiece of the second biennial *Report of the Fish and Game Warden* of Idaho summarized this view:

Laws for the protection of fish and game, if enacted with judgment and wisely administered, are essentially for the use and benefit of the middle and poorer classes. This must be conceded. The sportsman who is well supplied with this world's goods, if game be scarce in his immediate locality, has but to go where it is more plentiful and to his liking. The matter of expense is of no moment to him, while the man not so well favored, but with the sporting instincts just as strong and frequently better developed must of necessity content himself nearer home, and if the game and fish have not been properly protected forego the pleasure and delights of an outing entirely.[45]

Landowners, especially in rural communities, remained skeptical of the "official" position. The Colorado state fish and game commissioner reported that setters in remote areas "feel that they have a right existing from time immemorial to take what game they please for their own use and for that of their friends. . . . [A] great many take advantage of our inability to detect them in their violations." Even when caught, a jury of their own countrymen tended to excuse violations, "while if the offense was committed by a stranger in the land, or one coming from the city or town, the same jurors are only too glad to convict him for coming down and killing 'their' game."[46] A similar difficulty was encountered by the Wisconsin game warden who found district attorneys unwilling to enforce the law and justices of the peace who "usurped the prerogative of the supreme court, by declaring the fish and game laws unconstitutional and refused to hear the cases brought before them." Others freed defendants in spite of guilty verdicts.[47]

The serious difficulties in enforcing the game law hardly dampened the movement for reform in the management of wildlife. The state game law was clearly the wave of the future, concern for the "common man" for whom it is "almost impossible . . . to take up his rod or gun and leave his own dooryard without at the same time violating some statute,"[48] notwithstanding. But the idea of the game law is one thing; its realization is quite another. Great freedoms and small fortunes would be won and lost according to the form of the law and its mode of enforcement. Consequently, the parties to this elaborate dispute were hardly indifferent to the outcome. The next chapters trace their efforts to direct the policy process through the involvement of the courts, the legislatures, the state fish and game commissions, and by the manipulation of public opinion.

NOTES

1. See Ernest Staples Osgood, *The Day of the Cattleman* (Chicago: University of Chicago Press, 1957, reprint of 1929 edition).

2. John Bright, in an address to tenant farmers, offered this view of their burden:

A farmer becomes the tenant of certain lands which are to be the basis of his future operations, and the foundation on that degree of prosperity to which he may attain. To secure success, it is needful that capital should be invested, and industry and skill exercised . . . but the seed you sow is eaten by the pheasants; your young growing grain is bitten down by the hares and rabbits, and your ripening crops are trampled and injured by a live stock which yields you no return, and which you cannot take to market.

You are subjected to the incessant watching of the army of gamekeepers who patrol your fields, to see that neither yourselves nor your labourers interfere with the sports, nor the subsistence, of the lives of the sacred animals which the law dignifies with the name of game. In many cases you cannot keep a dog or a cat about your premises, and you cannot carry a gun across the fields for the occupation of which you pay a heavy rent, without exciting the suspicion of the ever-watchful gamekeeper whom your landlord employs to dog your very footsteps. That these things are irritating to you is clear. . . . [Richard Griffiths Welford, *The Influences of the Game Laws; Being Classified Extracts from the Evidence Taken before a Select Committee of the House of Commons on the Game Laws* (London: R. Groonbridge and Sons, 1846), pp. 5-7].

3. *American Sportsman* 4 (April 4, 1874), 10.
4. *Forest and Stream* 7 (December 14, 1876), 297.
5. *American Sportsman* 4 (June 6, 1874), 155-56.
6. Quoted in *Forest and Stream* 5 (January 13, 1876), 361.
7. *American Sportsman* 4 (May 9, 1874), 88. More sophisticated, but not widely held positions, saw property in animals and in land as constrained by considerations of social welfare. Thus, "A man's property is simply a trust that he holds for the benefit of the entire people" [ibid., 3 (October 25, 1873), 56] and "the free use of wild animals . . . [is] qualified by a regard for the rights of our neighbor and successor" [ibid., 4 (August 29, 1874), 344].
8. *American Sportsman* 7 (October 2, 1875), 8.
9. *American Sportsman* 6 (May 5, 1875), 72.
10. *Forest and Stream* 37 (December 17, 1891), 429.
11. *The New York Times*, October 22, 1893, p. 24.
12. The landowner's increased awareness of the value of wildlife on the farm is the subject of Chapter 3, text at notes 58-69.
13. U.S. Department of Agriculture, *Yearbook, 1904*, "Some Bene-

fits the Farmer May Derive from Game Protection," by T. S. Palmer (Washington, D. C.: G.P.O., 1905), p. 516.

14. *American Sportsman* 9 (March 31, 1877), 408.

15. *American Sportsman* 4 (July 4, 1874), 216.

16. *Forest and Stream* 15 (November 25, 1880), 323-24. One offer by sportsmen involved the payment of twenty to twenty-five cents for each quail killed. There is no indication of the numbers of landowners who had access to the journal.

17. Bradley, "Preserving Game," *Macmillan's Magazine* 58 (1888), 365. The leasing of sporting grounds is more common in recent years. A 1962 summary shows that in Wisconsin and Florida goose-hunting lands lease for as much as ten dollars per acre. Prime lands in Missouri have been leased for $1,000 per acre, and single shooting pits for $150-500 per year. A survey of 157 hunting clubs in California indicated that they leased an average of 3,280 acres at .36 per acre. Missouri deer lands leased for an average of .37 per acre, and those in New York for .22 an acre. Texas, with little public land for shooting, sees range lands near Houston leased at up to $5 per acre [Arnold W. Bolle and Richard D. Taber, "Economic Aspects of Wildlife Abundance on Private Lands," *Transactions of the North American Wildlife Conference* (Washington, D.C.: American Wildlife Institute, 1962), pp. 255-67].

18. *Forest and Stream* 1 (January 22, 1874), 377.

19. *Forest and Stream* 2 (May 14, 1874), 210.

20. Maine, *Report of the Commissioners of Inland Fisheries and Game,* 1883, pp. 18-19.

21. U.S. Department of Agriculture *Yearbook, 1904,* p. 512. One suspects that the views of the market hunter held by the sporting gentleman were not far different from the views of the poacher held by English nobility. "A poacher generally exhibits external marks or characteristics of his profession: the suspicious leer of his hollow and sunken eyes, his pallid cheek, his wide, copious and well-pocketed jacket—in fact, his appearance altogether is impressively at variance with that which is manifested by any other class of the human species" [T. B. Johnson, *The Gamekeeper's Directory, and Complete Vermin Destroyer* (London: Sherwood, Gilbert, and Piper, 1832), p. 138].

22. Notice should be taken of a project begun by Charles Hallock, while he was editor of *Forest and Stream,* to establish a community of sportsmen-farmers, a combination which he felt was highly compatible. He envisioned a town including 4,000 areas of fertile prairie and 400 town lots. The disposition of the property was to be made with

considerable care so as to gather a group "mindful of each other's prerogatives, and united in a common policy for the welfare of the town and the protection of the abundant game around about, which was the first desideratum sought in making a selection of a town site" [*Forest and Stream* 13 (August 28, 1879), 590]. Apparently Hallock never progressed very far with his idea. The town, still existing in northwest Minnesota and bearing its founder's name, was laid out in December 1879 on the south half of section 12 of the township following the decision of the St. Paul and Pacific Railway to locate its depot and sidings there. When James J. Hill took over the railroad, plans changed and the new townsite was platted on the northwest quarter of section 12. Hallock, unaware of the change, purchased half interest in the old townsite. Having discovered his error he added to this a five-acre plot adjoining the new site. Here he erected a large hotel in 1880, which soon became the center of town as well as the headquarters for many of his eastern sporting cronies. In 1892, the hotel, uninsured, burned to the ground. The rest of his scheme never materialized [*History of the Red River Valley*, 2 vols. (Chicago: C. F. Cooper & Co., 1909), pp. 937-90; Charles Hallock, *An Angler's Reminiscences: A Record of Sport, Travel and Adventure, with Autobiography of the Author* (Cincinnati: Sportsman's Review Publishing Co., 1913), p. 4].

23. See, for example, Wakeman Holberton, "The Supply of Game: Influence of Clubs and Private Game Preserves," *Harper's Weekly* 37 (April 8, 1893), 339.

24. Quoted in Phillip O. Foss, ed., *Conservation in the United States: A Documentary History-Recreation* (New York: Chelsea House Publishers, 1971), p. 8.

25. *American Sportsman* 1 (December 1871), 2; ibid., 1 (March 1872), 6.

26. Charles Hallock, *The Fishing Tourist: Angler's Guide and Reference Book* (New York: Harper & Brothers, 1873), p. 227.

27. Reprinted in *American Sportsman* 3 (November 15, 1873), 100.

28. New York: W. S. Rossiter, 1901. The association celebrated its centennial in 1971 [Theodore Whaley Cart, "The Struggle for Wildlife Protection in the United States, 1870-1900: Attitudes and Events Leading to the Lacey Act" (Ph.D. diss., University of North Carolina, 1971), p. 92, n. 31].

29. *Forest and Stream* 41 (August 19, 1893), 143. These private preserves served also as natural laboratories for scientific forest management. The Adirondack League Club's purchase, in particular, attracted the attention of B. E. Fernow, chief of the Division of Forestry of the Department of Agriculture, who, at the request of the club, prepared a

report on the potential for forest management. The chief recommended the installation of a network of permanent roads, a rail line through the property, and the development of nearby manufactures, such as pulp mills, which would use inferior timber and refuse. These changes are not entirely consistent with the maintenance of pristine hunting grounds [U.S. Department of Agriculture *Report of the Secretary of Agriculture*, 1890 (Washington, D.C.: G.P.O, 1890), pp. 214-23].

30. *Forest and Stream* 43 (December 22, 1894), 563.

31. *Forest and Stream* 42 (March 10, 1894), 199.

32. U.S. Department of Agriculture, Biological Survey Circular No. 72, "Private Game Preserves and Their Future in the United States" (Washington, D.C.: G.P.O., May 4, 1910), pp. 7-8.

33. Alexander Hunter described some of the clubs on the Sound. The oldest was the Swan Island Club, with a membership fee of $5,000. The Pamunkey Club, located on a four-acre island in the Sound, had four members (three from New York and one from Boston) and required a fee of $1,800. Others were the Lighthouse Club, with twenty members and a fee of $1,000, and the Currituck Inlet Club with thirty members and a fee of $300. All had additional annual charges ["The Club-Houses of Currituck Sound," *Harper's Weekly* 36 (March 12, 1892), 253-55].

34. For a description of the clubs in the St. Clair Flats region, see the series by E. Hough in *Forest and Stream* 35 (August 28, 1890), 112; (September 11, 1890), 151; (September 18, 1890), 171; (September 25, 1890), 191; (October 2, 1890), 211; (October 9, 1890), 231; (October 23, 1890), 273.

35. *Forest and Stream* 16 (April 28, 1881), 243. Later in 1881, the club was reported to control some 240,000 acres and 170 miles of shoreline [ibid., 16 (July 28, 1881), 514]. Worthy of mention are the private game preserves, among the largest of which was maintained in New Hampshire by Austin Corbin. His park of 27,000 acres was surrounded by a high fence which opened at nine gates, each with a keeper's cottage. *Forest and Stream* estimated his outlay at $400,000. The initial stock of wildlife of 25 bison, 60 elk, 70 deer, 6 caribou, 6 antelope, 18 boar, and 12 moose had, by 1895, become 750 elk, 700 deer, 500 boar, 100 moose, and 55 bison [ibid., 36 (March 13, 1891), 148-49, and 44 (May 11, 1895), 364-65].

36. *Forest and Stream* 16 (July 7, 1881), 453.

37. Henry Chase, *Game Protection and Propagation of America* (Philadelphia: J. B. Lippincott Company, 1913), p. 163.

38. *The Wilderness Hunter* (New York: G. P. Putnam's Sons, 1893), p. 449. In England, too, there was criticism. *Punch* suggested a series of new hunting regulations:

1. The Birds should be allowed to leave their cages before they are made the marks of the Sportsman.
2. No Sportsman should fire at a Bird with a gun having more than four barrels, unless he gives it (the Bird) a clear start of three yards.
3. Birds answering to pet names should be allowed to see the gun of the Sportsman before they are fired at.
4. A Bird settling on the shoulder of a Sportsman not should be fired at until it (the Bird) rises to fly away.
5. Birds should not be chained by their legs to the trees, unless they (the Birds) are very wild, and show a decided disposition to fly away [quoted in *American Sportsman* 5 (January 1, 1875), 279].

39. George Catlin, *Letters and Notes on the Manners, Customs, and Condition of the American Indians,* 2 vols. (1841; Minneapolis: Ross and Haines, 1965), I, 261-62.

40. Over 350,000,000 acres of public domain were disposed of between 1870 and 1900 under various land acts, or by military warrant or agricultural college scrip, or were selected by states and railroads. In 1900, over 900,000,000 acres remained unappropriated and unreserved, nearly two-thirds of it unsurveyed (U.S. *Statistical Abstracts,* various years).

41. The first federal reservation for the protection of wildlife was actually the Afognak Forest and Fish Culture Reserve in Alaska, created by President Harrison in 1892. Designed originally for the protection of salmon spawning grounds, the boundaries were extended, at the suggestion of Grinnell and others, to include some marine mammal habitats [James B. Trefethen, *An American Crusade for Wildlife* (New York: Winchester Press, 1975), pp. 123-24].

42. U.S. Department of Agriculture, *The Report of the Commissioner of Agriculture for the Year 1864,* "The 'Game Birds' of the United States," by D. G. Elliot (Washington, D.C.: G.P.O., 1864), p. 381.

43. Robinson, *Facts for Farmers,* 2 vols. (New York: A. J. Johnson, 1866) I, 198.

44. *Forest and Stream* 17 (January 26, 1882), 503.

45. Idaho, *Report of the Fish and Game Warden*, 1907/08, p. 4.

46. Colorado, *Report of the State Game and Fish Commissioner*, 1904/06, p. 11.

47. Wisconsin, *Report of the Commissioners of Fisheries and State Fish and Game Warden*, 1893/94, p. 30. Tales of this kind are common in the reports of those charged with the enforcement of the game laws. A California warden arrested two men for illegally netting salmon. One of the two admitted to the charge in court, and the judge instructed the jury that the admitted activity warranted a guilty verdict. After a ten-minute deliberation, the jury found him not guilty. "From the court-room to the hotel, and from the hotel to the station, we were followed by a howling mob of thirty or forty men and boys. It was a most insulting demonstration" [*Biennial Report of the State Board of Fish Commissioners of the State of California for the Years 1891/92* (Sacramento, 1892), p. 18].

48. Chase, *Game Protection*, pp. 151-52.

5

THE GAME LAWS
AND THE COURTS

Our laws upon the subject of game, and fish, seem to be a
heterogeneous mass of special enactments, passed at the
suggestion of various members of past legislatures. They
are empirical, and seem to have no coherence or general
design; and if carried out it would be difficult to say whether
they would be a benefit or harm to the game and fish of the
State.

First Annual Report of the State
Fish Commissioners of Minnesota, 1875

The Minnesota fish commissioners spoke the truth. The game
laws were not simple, nor uniform, nor stable, nor enforced,
nor were they motivated by any well-formulated management
goals which linked regulations to resulting wildlife populations. Many restrictions, such as those limiting open seasons,
controlling hunting methods, protecting females, and offering
special access to landowners, were carried over from colonial
and even from English statutes. Others, such as those setting
bag limits, controlling the export and import of game, and
requiring hunting licenses, grew in popularity toward the end
of the nineteenth century as game became increasingly scarce,
as states explored the constitutional limits of their control, and
as a commitment to enforcement called for revenues.

Beginning about 1870 the ideology of protectionism moved
throughout the nation, from the northeast, across the north to
the west, and gradually into the south. At each stop it confronted local custom, existing statute and common law, state
legislatures, the courts, sportsmen, market hunters, farmers,
and the U.S. Constitution. As protectionism became well entrenched in a particular region or with respect to a particular

kind of hunting behavior, it left behind an organized and committed lobby which would work to spread the gospel in areas still dominated by sentiments of free access and local control over wildlife. Early game laws, whatever their ultimate merits as instruments of wildlife protection, were hardly frivolous exercises of legislative authority. They were introduced, debated, amended, and enacted in the interest of certain legislators or in the interest of those with legislative attention. It is worth a brief excursion into the statute books of the period around 1870 to discover the great diversity of game laws with which late-nineteenth-century interest groups had to contend.[1]

THE LAW IN 1870

The northeastern hunter of 1870 faced the oldest and most comprehensive, though not most frequently enforced, game laws in the nation. Many species, especially deer and game birds, were routinely protected by seasons closed to hunting. Some states restricted efficient hunting technologies such as hounding deer or trapping, snaring, and netting upland birds.[2] Several had already extended protection to nongame species.[3]

In the absence of enforcement, the convention by which the dates of the open season coincided with the dates of legal sale and possession of captured game caused no difficulties either for hunters or for marketmen. By the early 1870s, particularly in New York City but also in other major urban market centers, law enforcement by game protection associations called forth demands for legislated marketing periods that extended beyond the hunting season.[4] Consequently, the 1871 New York law (Ch. 721, pp. 1669-81) offered the two months following the close of quail and grouse season for the disposal of legally killed birds. The enforcement problems implied by this approach are enormous. Once a bird is dressed for sale or frozen for storage, its time and place of death are very difficult to determine. This trade off between the ease of enforcement and the disposal of lawfully captured game was to characterize much of the conflict between sportsmen and marketmen well into the twentieth century.

The game laws also ran afoul of the growing interest in natural history, accompanied by the collection of wildlife, and especially bird specimens, for private natural history cabinets and public museums. Early laws merely exempted such activities from control; as collectors increased in number and as the feather and egg trades encouraged the use of such exemptions as blinds behind which to collect for the market, states began to issue collecting permits for scientific and educational purposes. But even bona fide collectors posed real threats to scarce species as they competed for remaining examples of particular life forms.

Dense occurrences of wildlife, such as wintering populations of waterfowl along the Chesapeake Bay and Currituck Sound, drew early protective attention and experimentation with forms of control not applied more widely until the turn of the century. An 1872 Maryland law (Ch. 54, pp. 74-81) prohibited the use of vessels within one-half mile of shore, prohibited the use of the huge, cannon-like punt and swivel guns, and limited the hunting of waterfowl to the daylight hours of Mondays, Wednesdays, and Fridays.[5] License fees of twenty dollars and five dollars per year were payable, respectively, by users of sink boxes and sneak boats. To enforce these and other provisions, an 1870 statute (Ch. 296, p. 507) ordered the commanding officer of the state Oyster Police to detail inspectors "at such times as he shall be able to spare them" during the wildfowl season.

North Carolina's laws developed along similar lines. An 1870 statute (Ch. XLII, p. 85) prohibited Sunday and night hunting and the use of any gun "other than can be fired from the shoulder," as well as the construction of any blinds, boxes, or batteries in any waters away from the shore. Perhaps in an effort to attract tourists while retaining control over market hunters, the act was supplemented in 1871 (Ch. C, p. 163) to exempt "such non-residents who resort to the waters of Currituck Sound for the sole purpose of shooting game as sportsmen, and who shoot over or on land or marshes owned or leased by them, and who do not kill game for a foreign market." At this date, there were a number of wealthy northerners who, as members of North Carolina shooting clubs, would have had an interest in the passage of a provisions such as this.[6]

A significant characteristic of early game legislation, particularly in states and territories of the far west, was the distinction between the utilitarian needs of the frontiersman and local resident and the perceived decadence of the eastern sportsmen and trophy hunter. Oregon, in 1874 (pp. 6-8), prohibited the taking of deer or elk to obtain horns, hide, or hams as the sole prize, yet made big game available, without closed season, for personal consumption. In a similar vein, a Colorado statute of 1872 (pp. 134-36) required that the hunter of big game bring all edible portions to market.[7] The provisions of the Yellowstone Park Act, passed by Congress in 1872, indicated that there was sympathy for this view on a national level. The act directs the secretary of the interior to attend only to "the wanton destruction of the fish and game found within said park, and [to] their capture or destruction for the purposes of merchandise or profit."[8]

Only the Deep South lacked a statutory commitment to wildlife protection during this period. Alabama, for example, had no general game law until 1907. The prohibition of fire-hunting, enacted originally as part of a law relating to the Mississippi Territory in 1803, provided only a penalty (thirty-nine lashes) for violation by slaves. The law was renewed for the state in 1822 (p.57) as "An Act to Suppress the evil and pernicious practice of Fire-Hunting," and applied only to deer hunters in possession of gun and fire at night. The only game laws in force in 1870 applied to three counties (1854; No. 317, pp. 204-5). An 1858 Georgia law prohibited nonresidents from hunting and fishing within the limits of the state, but "as it was passed to prevent strangers and others from holding any conversation with slaves, its provisions are not applicable to the present state of affairs."[9] Except for the prohibition of Sunday hunting (1855, p. 180), Arkansas had no game law until 1875. Of this state and Texas, which had no general game law until 1903, *Fur, Fin, and Feather* printed the following remarks in place of its usual summary:

We are unable to state whether there are any laws in this State for the protection and preservation of game or not. We have endeavored to obtain the desired information from the State authorities, but have been unsuccessful in getting any light on the subject; and all the Statutes and laws of the State we have had access to are totally silent in regard to the matter. And we are therefore inclined to believe that Arkansas, like Texas, has no game-protecting laws, and that the pot-hunter, who is held in utter detestation by every legitimate sportsman, is here left to pursue his wanton destruction unmolested.[10]

The legal basis for these and other forms of regulation was hardly examined during this period. The state's proprietary interest in wildlife, the origins of which are considered below, was nonetheless implicit in various legislative pronouncements. The constitutional guarantees of liberty "in seasonable times, to hunt and fowl . . . under proper regulations" granted to the residents of Vermont and Pennsylvania have already been noted.[11] Apparently, the state might reserve wildlife for its own residents and might offer those residents the freedom to pursue game onto private lands. Alternatively, it might define such pursuit as trespass and offer special hunting privileges to landowners.

The Connecticut legislature, in 1861, (Ch. LXXI, pp. 102-3) carefully defined trespass as entering the lands of another for the purposes of hunting contrary to the provisions of the game law and without the consent of the owner or occupant. Presumably entering such lands, with or without permission, for the purposes of lawful hunting was an unquestioned right. Similarly, a landowner might extend an invitation to engage in otherwise unlawful hunting. But by 1871 the tide had begun to turn in favor of the landowner. A New York law that year (Ch.721, pp. 1669-81) offered protection to landowners who would assume the responsibility of posting their lands against trespassers.

Other discriminations in favor of landowners were widespread. Kentucky in 1861 (Ch. 309, pp. 42-43) and Delaware in 1871 (Ch. 58, pp. 69-70), for example, protected song and insectivorous birds except against the landowner on his own land. Extreme examples of landowner's rights may be found in special statutes relating to preserves and shooting clubs. The Blooming Grove Park Association, incorporated in Pennsylvania in 1871, was granted the right to write the game laws which would apply within its enclosed grounds, to hire game wardens (to the limit of one per hundred acres), who would be deputized by the county sheriff, and to impose penalties for violation.[12] Sorting out the rights of landowners versus non-owners and of state residents versus nonresidents remain among the thornier problems of wildlife law.

REGULATION AND THE COURTS

Occasional enforcement of these early laws invited court comment on the constitutionality of various restrictions which states placed on individual hunters. Between 1875, when the New York Court of Appeals upheld the power of the state to regulate the taking, sale, and possession of game, and 1896, when the U.S. Supreme Court issued its landmark decision establishing the states' proprietary interest in wild animals, the legal history of wildlife was rediscovered and thoroughly scrutinized for guiding principles.

The legal issues which arose were symptomatic of judicial controversies of the nineteenth century. First, the human population density increased and as activities productive of external effects became increasingly profitable, conflicts arose among individuals. With respect to wildlife, the right to free access to game ran head on into the rights to improve lands and to post property. More specialized conflicts faced two hunters in pursuit of the same animal; did the prize belong to him who started the game or to him who delivered the mortal blow?[13] A second controversy centered on the power of the state vis-à-vis the individual and examined the growing conflict between due process in the protection of private property and police power in the protection of general welfare.[14] Although the freedom of individual action was generally broadened during the post-Civil-War period, access to game was markedly narrowed through legislation designed to protect a common heritage in wildlife. Finally, with the evolution of a national economy and the proliferation of activities that spilled over state boundaries, what were to be the rights of states in a federal system? The increasing prominence of the federal profile in the late nineteenth century could not but have had wide-ranging implications for control over wildlife. Among the more obvious changes were those accompanying the reversal of the century-long process in which a major portion of the public domain was transferred into the hands of states and private parties. The creation of several national parks and the reservation of millions of acres of western forest lands forced a federal position on resource management. The broadening interpretation of the commerce clause of the Constitution as a basis for federal intervention would prove equally important. States' control over wildlife was weakened as the federal government sought to protect its own property and to control goods flowing among the states.

The development of wildlife law[15] as traced here suggests no search for an immutable body of legal truth, such as was said to exist in absolutist treatises of the eighteenth and nineteenth centuries, but rather a continual reinterpretation of legal doctrine in the context of a changing environment.[16] Thus, in wildlife law we find support for the state ownership doctrine in spite of its questionable historical basis, and we find expanded federal jurisdiction over wildlife when, at an earlier time, such jurisdiction was thought to overstep constitutional bounds. Supreme Court Justice Oliver Wendell Holmes, in the landmark decision of *Missouri* v. *Holland*, which upheld the Migratory Bird Treaty with Great Britain, observed that the Constitution has "called into life a being the development of which could not have been foreseen completely by the most gifted of its begetters. . . . The case before us must be considered in the light of our whole experience and not merely in that of what was said a hundred years ago."[17]

By English, and before it, Roman, common law, animals *ferae naturae* (of a wild nature) are *nullius bonis* (the property of no one) and are thus common property. The law recognizes certain forms of private property, each with its proper Latin name, which result from capture, confinement, or grant.[18] The game laws specify the limitations on the freedom of individuals to reduce animals *ferae naturae* to private property.

The English game laws, which formed the basis for colonial, and later, state and federal controls, originated following the loss of royal control over forested lands.[19] After the Norman invasion, existing forests and all game (but not all wildlife) were appropriated by the king as his private property. New royal forests were created by depopulating entire regions. Trespass and poaching were strictly punished, often by loss of life or limb. The forests were protected primarily as places where the king and those whom he designated as worthy might engage in the chase. The king might have favored barons and other persons of high social status by conferring upon them the franchise over game on their own property. As the power of the landed gentry grew, individual grants of privilege were superseded by general game laws that described the nature of access to game for the society as a whole.[20]

The particular version of this history that came to be accepted in the United States viewed the king as strict owner, by prerogative, of all lands and resources thereon, including game animals. Use of the land and access to its resources could be obtained only if the king vested his rights in another. Thus, just as the king was the owner of game in England, so the individual colonies and later the states, by the transfer of royal authority, owned the game found within their boundaries.

So much had long been assumed by colonial and state governments seeking to guarantee their citizens the nondiscriminatory access to game which had been blocked by European class structure and patterns of land ownership. Sir William

Blackstone's *Commentaries on the Laws of England,* popular in the colonies and new nation from their first publication in 1765-69, offered convenient documentation.[21] But much evidence questions this interpretation.[22] The difficulties were initially pointed out by Edward Christian in his *Treatise on the Game Laws,* published in 1817. If all wild game were indeed the unquestioned property of the king, why should particular species be singled out as royal species? He cited the *Case of the Swans* (1585) in which the English jurist Sir Edward Coke referred to swans as "royal fowl" much as "whales and sturgeons are royal fish, and belong to the king by his prerogative." Although the king may have claimed all wild animals as his property, his claims came to be limited to a few species by common law, just as his so-called right to afforest lands became limited to the right to afforest only demesne lands.[23]

As the forest laws gave way to common law and the increasing political power of the landed gentry, Parliament maintained the protection that the forest laws had provided by enacting specific game laws as early as 1390. These laws represented, in fact, a concentration of the rights in game. Whereas under complete royal ownership all landowners had, in effect, the rights to game on their own lands except as those lands were afforested by the king, under the series of laws which was enforced into the nineteenth century, certain property owners were stripped of that right.[24] As the wealth requirements for property in game diminished, the general principle emerged and was confirmed in *Blades* v. *Higgs* (1865) that the landowner, by virtue of his ownership of the soil, had a right to wild animals on or over his property.[25] While Parliament might have controlled access to wild animals, it was police power and not ownership that provided the authority.

Case law in the United States supports state intervention in the hunt both as an exercise of police power and of its proprietary interest in wild animals. The police power doctrine emerged clearly in the 1875 case of *Phelps* v. *Racey,* which concerned the illegal possession of quail by a game dealer in

New York City.[26] The New York Court of Appeals sustained the dealer's conviction and concluded:

The legislature may pass many laws the effect of which may be to impair or even destroy the right of property. Private interest must yield to the public advantage. . . . The protection and preservation of game has been secured by law in all civilized counties, and may be justified on many grounds, one of which is for purposes of food. The measures best adapted to this end are for the legislature to determine, and courts cannot review its discretion.[27]

The state ownership doctrine, with respect to terrestrial species, finds its earliest legal expression in an 1881 Illinois decision, upholding the prohibition of quail marketing during the closed season, in which the court thought it "accurate to say that the ownership of the sovereign authority is *in trust* for all the people of the State. . . ."[28] But elsewhere in the decision, regulation was supported by police power. The Supreme Court of Minnesota, in an 1894 decision, while questioning the right of the legislature "to declare all wild game the property of the state, in a proprietary sense," took it "to be the correct doctrine in this country that the ownership of wild animals, so far as they are capable of ownership, is in the state, not as proprietor, but in its sovereign capacity, as the representative, and for the benefit, of all its people in common."[29]

The U.S. Supreme Court had no occasion to comment directly on the source of state authority until 1896 in *Geer v. Conn.*[30] The *Geer* case involved the right of the state to prohibit the exportation of game taken legally within the state. While the basic authority of the state was seen to rest with its police power, the court, following a rather lengthy discussion of European and English law and their transfer to the colonies (as reported by Blackstone), asserted the state's proprietary interest in wild animals. Legislation of the kind contested here surely depended on the ownership doctrine for, in its absence, the state would be obstructing commerce by prohibiting the passage between states of private property.[31] Because wildlife is the property of the state, the court ruled, the state may determine the conditions under which its citizens might acquire property in game. The argument was succinctly put in Justice Edward D. White's decision:

The sole consequence of the provision forbidding the transportation of game, killed within the State, beyond the State, is to confine the use of such game to those who own it, the people of that State. The proposition that the State may not forbid carrying it beyond her limits involves, therefore, the contention that a State cannot allow its own people the enjoyment of the benefits of the property belonging to them in common, without at the same time permitting the citizens of other States to participate in that which they do not own.

That the provision is a violation of the interstate commerce clause is to presuppose

that where the killing of game and its sale within the State is allowed, it thereby becomes commerce in the legal meaning of that word. In view of the authority of the State to affix conditions to the killing and sale of game, predicated as is this power on the peculiar nature of such property and its common ownership by all the citizens of the State, it may well be doubted whether commerce is created by an authority given by a State to reduce game within its borders to possession, provided such game may not be taken, when killed, without the jurisdiction of the State. . . . The qualification which forbids its removal from the State necessarily entered into and formed part of every transaction on the subject, and deprived the mere sale or exchange of these articles of that element of freedom of contract and of full ownership which is an essential attribute of commerce.[32]

The prohibition of export of legally acquired wildlife was but one kind of state regulation that was thought by some to run afoul of the commerce clause of the U.S. Constitution. A second kind of regulation prohibited the importation of game legally acquired elsewhere. It seems apparent that by the *Geer* ruling game taken in a state that permits the export of legally acquired animals remains private property upon export and enters the stream of interstate commerce. Its movement among states may not be hampered by individual states. At the same time, states with depleted game covers would have a clear interest in such legislation. Game laws are considerably easier to enforce in the marketplace than in the field, and the flow of game from states with abundant wildlife into the marketplace provides a safe cover for the continued hunting of local populations. At least two state courts, in the interest of protecting these local populations, upheld the application of laws prohibiting possession and sale of imported game. The St. Louis Court of Appeals opined in 1876: "We see nothing unconstitutional in the act. The game law would be nugatory if, during the prohibited season, game could be imported from the neighboring States." The Supreme Court of Ohio upheld a similar ruling in 1894 by arguing that such a statute offers more protection to "birds and game in this state than one preventing the sale of such only as should be killed here."[33]

In these cases, the issue of importation is merely skirted, for nowhere is there incontrovertible evidence of importation. Game killed out of state has bearing on the populations of game within the state only insofar as it or its offspring might have migrated there at some time in the future if it had not been killed.[34] This was observed by the Supreme Court of Pennsylvania which, in 1891, noted that the object of the state game law "was the preservation of game within this commonwealth. ... The law was not intended to have any extraterritorial effect, and, if it was, it would be nugatory."[35]

The power to regulate interstate commerce lies with the federal government, and the protection that the states desired by legislation prohibiting the importation of game was provided by the Lacey Act of 1900 which declared, in part:

all dead bodies, or parts thereof, ... of any wild game animals, or game or song birds transported into any State or Territory, or remaining therein for use, consumption, sale, or storage therein, shall upon arrival in such State or Territory be subject to the operation and effect of the laws of such State or Territory enacted in the exercise of its police powers, to the same extent and in the same manner as though such animals or birds had been produced in such State or Territory, and shall not be exempt therefrom by reason of being introduced therein in original packages or otherwise.[36]

THE GAME LAW AT THE TURN OF THE CENTURY

It was against this treatment in the courts that the game laws of the several states evolved through the turn of the century. During this period, the number of states and territories offering seasonal protection to important species grew as did the extent of protection. Tables 2-5 illustrate the development of seasonal protection by comparing the periods closed to hunting of deer and ducks in 1871 with those periods in 1901.[37] But while the extent of seasonal protection increased greatly, there was no greater uniformity of seasons among states.[38] The importance of uniform seasons, both for easing the problems of enforcement and for maintaining populations themselves, provided major motivation for a series of national sportsmen's meetings convened during the 1870s and 1880s.[39] The continued failure of such efforts underlines one serious limitation of the state ownership doctrine. Migrating species were subject to constant hunting pressure as they moved through states, each of which sought to provide fair access for its own hunters. The overall hunting pressure entered only indirectly into the calculus of individual state legislatures. Recognition of such interrelationships gave states an interest in, but no control over, the game laws of their neighbors.

Uniformity was lacking not only among states but within them. States often delegated power over game resources to counties and towns and more often allowed local exemptions from state game laws. In mid-nineteenth-century rural America, when a journey to the county seat was a major outing and local governments represented for many the limits of familiarity and trust, this was both logical and necessary. Who would know better the condition of local game populations, and who would better be able to protect local hunters from limitations of

Table 2. DEER SEASONS, 1871

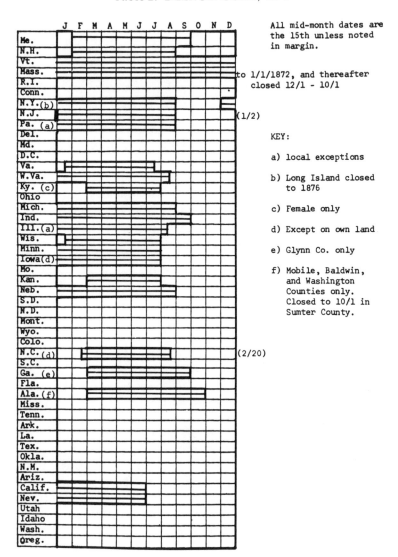

SOURCE: *Fur, Fin, and Feather: A Compilation of the Game Laws of the Different States and Provinces of the United States and Canada,* revised and corrected for 1871-72 (New York: M. B. Brown and Co., 1871); and statutes of individual states.

Table 3. DEER SEASONS, 1901

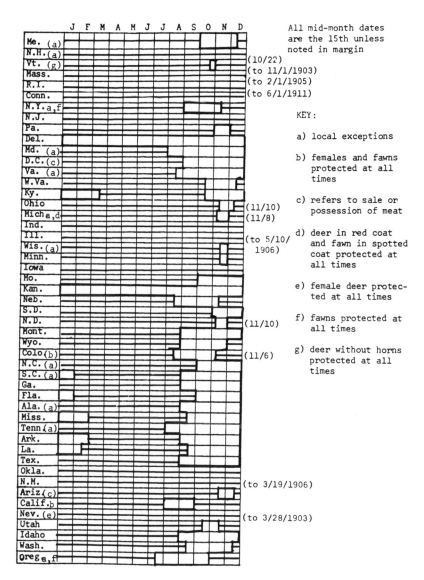

SOURCE: U.S. Department of Agriculture, Division of Biological Survey, Bulletin 16, *Digest of Game Laws for 1901*, by T. S. Palmer and H. W. Olds (Washington D.C.: G.P.O., 1901), Table I.

Table 4. DUCK SEASONS, 1871

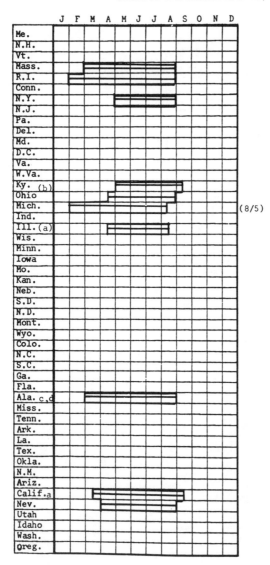

All mid-month dates are the 15th unless noted in margin

KEY:

a) local exceptions

b) except on own land

c) summer ducks only

d) Mobile, Baldwin and Washington Counties only

SOURCE: *Fur, Fin, and Feather: A Compilation of the Game Laws of the Different States and Provinces of the United States and Canada*, revised and corrected for 1871-72 (New York: M. B. Brown and Co., 1871); and statutes of individual states.

Table 5. DUCK SEASONS, 1901

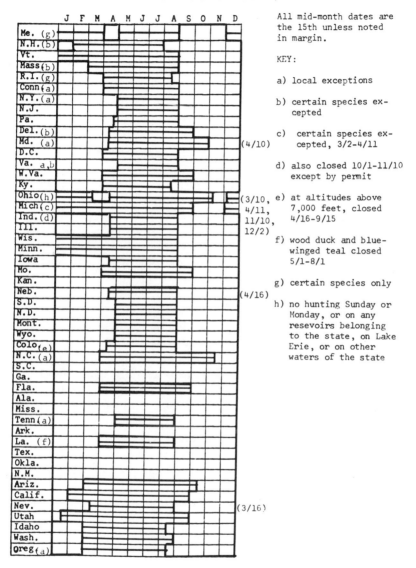

All mid-month dates are the 15th unless noted in margin.

KEY:

a) local exceptions

b) certain species excepted

c) certain species excepted, 3/2-4/11

d) also closed 10/1-11/10 except by permit

e) at altitudes above 7,000 feet, closed 4/16-9/15

f) wood duck and blue-winged teal closed 5/1-8/1

g) certain species only

h) no hunting Sunday or Monday, or on any reservoirs belonging to the state, on Lake Erie, or on other waters of the state

SOURCE: U.S. Department of Agriculture, Division of Biological Survey, Bulletin 16, *Digest of Game Laws for 1901*, Plate II and Table I.

155

freedom imposed by legislators from afar? These local provisions, which characterized the laws of most states, faded in the north by the last quarter of the century[40] but persisted elsewhere, particularly in the mid-Atlantic region.[41] The county laws in force in 1901 for Virginia, Maryland, and North Carolina showed almost as much variation as did the laws in widely separated states. To cite but one example, Virginia quail were hunted in seventeen seasons, beginning between October 1 and November 15, and ending between December 25 and March 1.[42]

Increased seasonal protection was but one direction of change in the game laws. Others represented the implementation of new regulatory approaches.[43] The few states that prohibited the sale or export of game before 1870 acted largely with respect to illegally captured animals. By 1901, a number of states prohibited the sale and/or export of a wide range of legally captured species. All states but five prohibited the export of at least one species, and twenty-nine states limited in some ways the sale of game. The comprehensive no-sale laws of Nevada and Michigan, including both native and imported game, were supported by the Lacey Act of 1900.[44]

These controls on the distribution of captured game were augmented by regulations limiting capture in the field. As shown on Map 2, twenty-four states set bag limits on one or more species. Map 3 indicates the distribution of states requiring hunting licenses, and their respective fees. The controversy generated by consideration of license laws is outlined in the following chapter. The license was later to provide a means for enforcing the bag limit, as license coupons could be affixed to captured big game.

All of these new forms of legislation, were they to be seriously regarded, required funding and a mechanism for translating funding into the protection of wildlife. The logical mechanism, a model for which already existed in some states in the fish commission, was a state-level administrative agency that might develop the necessary scientific expertise for wildlife management, enforce existing law by deploying agents in the field, and advocate the cause of wildlife before the legislature and the public. Map 4 illustrates the growth of state-level

Map 2. BAG LIMITS, 1901

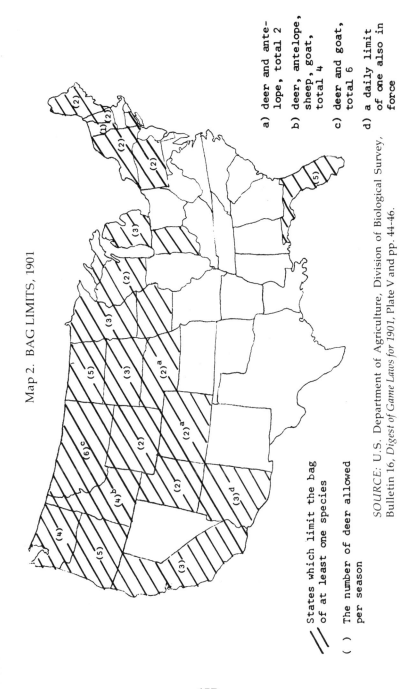

// States which limit the bag of at least one species

( ) The number of deer allowed per season

a) deer and ante-
    lope, total 2

b) deer, antelope,
    sheep, goat,
    total 4

c) deer and goat,
    total 6

d) a daily limit
    of one also in
    force

SOURCE: U.S. Department of Agriculture, Division of Biological Survey, Bulletin 16, *Digest of Game Laws for 1901*, Plate V and pp. 44-46.

Map 3. HUNTING LICENSES, 1901

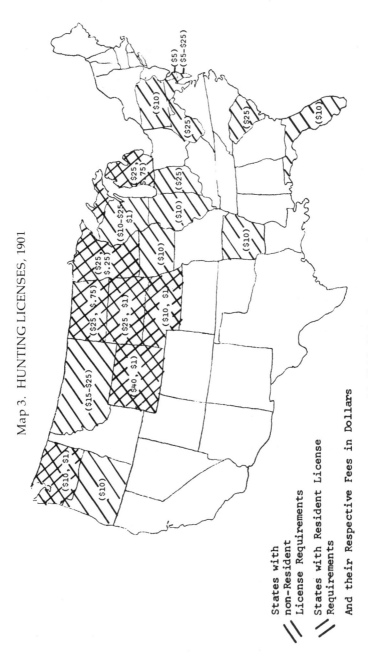

States with
non-Resident
License Requirements

States with Resident License
Requirements

And their Respective Fees in Dollars

($5)
($5–$25)

($10)
($25)
($25)
($10)
($25)
($.75)
($25)
($25)
($10–$25)
($1)
($10)
($25)
($.25)
($10)
($10)
($25, $.75)
($25, $1)
($10, $1)
($1)
($15–$25)
($40, $1)
($10, $1)
($10)

SOURCE: U.S. Department of Agriculture, Division of Biological Survey,
Bulletin 16, *Digest of Game Laws for 1901*, Plate VI and pp. 46–50.

fish and game commissions or wardens through the turn of the century. In many states, an existing fish commission was expanded to include responsibility for game resources. In some, a separate game agency was created, later to be joined to an existing fish commission.[45] As will be seen in the following chapter, the success and political independence of the several state agencies varied considerably; but whatever their early success, these agencies provided the backbone upon which present state level wildlife management is constructed.

INTERPRETATIONS

Although the present text is primarily concerned with the nineteenth century, the more recent legal history of wildlife merits some attention here.

The Lacey Act had the immediate effect of broadening state power in regulating wildlife, but its ultimate effect, by introducing the federal government as a competing regulator, was to make that power narrower. Since 1900, the scope of the ownership doctrine has been further narrowed. Twentieth-century interpretations suggest a wide role for federal activity in the control of access to wildlife resources. This appears to be the result both of the inability of legislation supported by nineteenth-century interpretations to protect wildlife to the extent felt necessary by conservationist interests and by the court, and of the changing court views in related matters that provide precedent for federal activity in wildlife regulation.[46]

The difficulties with the state ownership doctrine were particularly evident in regulating access to migratory species, especially birds, which range across the boundaries not only of states but of nations. The first attempt to place migratory animals under the control of the federal government was made in the House of Representatives by George Shiras, III, of Pennsylvania in December 1904. His bill sought to place all "migratory game birds which do not remain permanently the entire year within the borders of any State or Territory . . . within the custody and protection of the Government of the United States. . . . "[47] In spite of support by President Roosevelt, *Forest*

Map 4. THE DEVELOPMENT OF STATE-LEVEL FISH AND GAME
COMMISSIONS OR WARDENS, THROUGH 1901

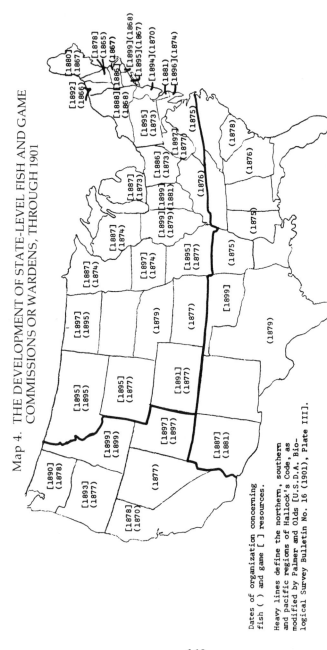

Dates of organization concerning
fish ( ) and game [ ] resources.

Heavy lines define the northern, southern
and pacific regions of Hallock's Code, as
modified by Palmer and Olds [U.S.D.A. Biological Survey Bulletin No. 16 (1901), Plate III].

SOURCE: U.S. Department of Agriculture, Bureau of Biological Survey, Bulletin 41, *Chronology and Index of the More Important Events in American Game
Protection, 1776-1911*, by T. S. Palmer (Washington, D.C.: G.P.O., 1912); and
statutes of individual states.

160

and Stream, and the Boone and Crockett Club, the measure failed to emerge from committee owing to the general view that it was unconstitutional. Over the next few years, a number of bills were brought before Congress by Shiras and others, and in March 1913, the desired legislation was engineered through Congress in the form of the Weeks-McLean Migratory Bird Act.[48]

The law was held to be unconstitutional by two federal district courts.[49] In each case, the court upheld state claims that migratory birds were owned by the states in which they were found. An appeal, taken to the Supreme Court, was dismissed with the passage of the Migratory Bird Treaty Act. This 1918 act, virtually identical to the statute which it replaced, gave effect to the 1916 treaty concluded between the United States and Great Britain. The 1918 act was upheld by four federal district courts and by the Supreme Court, in *Missouri* v. *Holland*, with reference to the supremacy clause of the constitution. Boyd observes that, significantly, the court "neither held nor intimated that Congress could not, absent a treaty, have passed the legislation involved in the case."[50] Justice Holmes took the occasion of this case to remark on the ownership doctrine:

The State as we have intimated founds its claim of exclusive authority upon an assertion of title to migratory birds, an assertion that is embodied in statute. No doubt it is true that as between a State and its inhabitants the State may regulate the killing and sale of such birds, but it does not follow that its authority is exclusive of paramount powers. To put the claim of the State upon title is to lean upon a slender reed. Wild birds are not in the possession of anyone; and possession is the beginning of ownership. The whole foundation of the State's rights is the presence within their jurisdiction of birds that yesterday had not arrived, tomorrow may be in another State and in a week a thousand miles away.[51]

The Supreme Court had no occasion to remark further on the ownership doctrine until 1948. In that year, two cases concerning the discriminatory access to shrimp in North Carolina and to fish in California provided the opportunity, though not the necessity, for comment.[52] Both decisions weakened the authority of the state. In the former, Chief Justice Fred Vinson issued what commentators Nicholas V. Olds and Harold W. Glassen refer to as the "death knell of the 'state ownership' doctrine."[53]

> The whole ownership theory, in fact, is now regarded as but a fiction expressive in legal shorthand of the importance to its people that a State have power to preserve and regulate the exploitation of an important resource.

The influence of the *Geer* decision continued to decline and, in the 1979 case of *Hughes* v. *Oklahoma*, the U.S. Supreme Court, noting that *Geer* had been eroded "to the point of virtual extinction," directly overruled the state's authority to prevent legally captured wildlife from entering the stream of interstate commerce.[54]

The federal role in wildlife protection finds support not only in this erosion of the state ownership doctrine but also in the powers enumerated in the Constitution.[55] In addition to the supremacy clause of Article VI already remarked upon, Article IV, section 3, conveys to Congress "the Power to dispose of and make all needful Rules and Regulations respecting the Territory or other Property belonging to the United States." The authority of the federal government to regulate access to wildlife on public land is well established and has been broadened to include the regulation of access on private land where such access imperils federal property.[56] A third enumerated power is found in the commerce clause of Article I, section 8. The meaning of "commerce" has grown increasingly broad. The recent inclination of the courts "has been to uphold federal regulation of anything passing between states without inquiry as to its commercial nature."[57] Thus, federal regulations cover the passage between states of lottery tickets, stolen automobiles, electrical impulses, air pollutants, and individuals with the intent to incite riot.[58] Wildlife, in passing across state lines, similarly might be regulated.[59] In addition, by the "affectation theory," Congress may regulate a purely intrastate activity that has an adverse effect on interstate commerce. Because racial discrimination in restaurants that served only local patrons placed a substantial burden on interstate commerce in food items, the Supreme Court upheld the public accommodations section of the 1964 Civil Rights Act.[60] The impact of hunters and fisherman on interstate commerce through their purchase of guns, ammunition, and other equipment is substantial.[61]

It appears, therefore, that the doctrine of state ownership of wildlife is transitory. It was not derived from English common law nor is it protected from encroachment by federal regulatory activity. It was, however, viewed as necessary to justify certain components of state regulation that had emerged by the end of the nineteenth century.

One might wish to demonstrate that the configuration of property rights that did emerge during this period was in some sense optimal and that therefore the "correct" interpretation was made by the courts. But the notion of "correctness" in history is elusive. That it was "correct" in terms of the demands of the various interest groups concerned is open to little question if the structure of demand is measured in retrospect by changes in public policy. But this is mere tautology. On the other hand, to ask whether the institutionalized forms that emerged in the nineteenth century were "correct" with respect to demands that were to emerge in the twentieth century is somewhat meaningless, particularly in that twentieth-century demands are predicated on nineteenth-century events.

It is more manageable to comment on the pattern of wildlife abundance and distribution that might have emerged in the absence of the doctrine of state ownership. The powers that the states had assumed through 1900 included the regulation of the time and place of the hunt, of the sale of game, of its export and import, and of the behavior of nonresidents. Were the state deprived of ownership as justification for its regulation and forced to rely only on the doctrine of police power, the regulation of seasons, methods, and times of hunt could continue to be justified. It seems likely, however, that the prohibition of export would not have been upheld. Further, the attempts of the state to prohibit the importation of game legally taken in other states were generally held invalid, irrespective of the doctrine of ownership. Finally, by the privileges and immunities clause, substantial discriminations against nonresidents may not have been upheld.[62]

Had there been no proprietary interest on the part of the state, it seems clear that wildlife populations would have been hunted more intensively before 1900. Those aspects of protection that would not have existed were those tending to preserve the wildlife in a given state for its own residents. The prohibitions of export tended to keep game from being shipped to market in states in which native populations of wildlife were small relative to the demand for killed game and in which no prohibition of sale existed. There is no reason to believe that such states would have passed laws forbidding sale were export prohibitions considered invalid. On the contrary, their own citizens stood to gain by keeping the market price low with game imported from distant states. By discouraging some hunters from selecting an otherwise favored state, nonresident license requirements also reduced the demand on local wildlife populations.

Whether, in the absence of the state ownership doctrine, protective regulation would have emerged on other governmental levels is highly speculative. Regulation by individual towns and cities can be no greater than that allowed by the state and, therefore, can be no more effective. Further, the high costs of organizing cooperative efforts militate against effective control on this level. The federal government, on the other hand, was only beginning to exercise its authority in regulating economic activity by the end of the nineteenth century. It seems unlikely that the increased need for regulation in a single area would have altered the speed of this development.

In conclusion, an important virtue of the state ownership doctrine was its ability to provide a legal justification for certain regulatory powers that several states had assumed in order to limit access to wildlife resources within their borders. As has been indicated by the above summary of events, however, the scope of the doctrine has been increasingly narrowed in the twentieth century as it has become clear that the effective management of wildlife populations cannot be maintained if extensive rights of ownership lie with the state.

NOTES

1. Two useful compilations of the law are *Fur, Fin, and Feather, A Compilation of the Game Laws of the Different States and Provinces of the United States and Canada: To Which Is Added a List of Hunting and Fishing Localities, and Other Useful Information for Gunners and Anglers* (New York: M. B. Brown and Co., 1871), and U.S. Department of Agriculture, Biological Survey Bulletin No. 41, "Chronology and Index of the More Important Events in American Game Protection, 1776-1911" by T. S. Palmer (Washington, D.C.: G.P.O., 1912). See also, U.S. Department of Agriculture, *The Report of the Commissioner of Agriculture for the Year 1864*, "Birds and Bird Laws," by J. R. Dodge (Washington, D.C.: G.P.O., 1864). Certain inaccuracies in and information missing from these sources have led the author to check, when possible, the session laws of the individual states. Where this has been done, the act or statute number and/or the page number in the session laws follows, in parentheses in the text, the date of the law. Other sources are as noted.

2. Economic incentives worked against the enforcement of these controls on taking birds. Shooting, in addition to raising the cost of the harvest, left the consumer with the task of picking shot from his teeth; consequently, gunned birds brought a lower price in the market. An 1869 Pennsylvania law (No. 60, pp. 84-86) provided for the taking of quail by trap or net from November 5 to January 5, during which time they might not otherwise be captured, "for the sole purpose of preserving them alive over the winter." A similar clause appears in various forms in the laws of a number of states. Although it may have value in and of itself, because quail are particularly subject to winter kill in the north, the enforcement of the clause is clearly difficult.

3. In 1855, Massachusetts (Chap. 197, pp. 615-16) provided year-around protection for some song and insectivorous birds. Vermont, in 1851 (No. 30, p. 30), and Connecticut, in 1850 (Ch. V, p. 5), provided similar protection, except that the landowner or tenant was not constrained on his own premises. This discriminatory feature was removed

in 1861 in Connecticut (Ch. LXXI, pp. 102-3) and in 1878 in Vermont (No. 87, pp. 85-6).

4. On the politics of the New York game law and its enforcement by the New York (City) Association for the Protection of Game, see Chapter 6, text at notes 143-46.

5. This was but a formalization of agreements already in force among market hunters in the region. Their purpose was to increase, not decrease, the current harvest by encouraging birds to remain in an area accessible to hunters. Thomas Gilbert Pearson, *Adventures in Bird Protection: An Autobiography* (New York: D. Appleton-Century Company, 1937), p. 122.

6. Shooting preserves are discussed at length in Chapter 4, text at notes 22-35.

7. See also the statutes of Washington (1865 pp. 37-38) and Wyoming (1871 pp. 113-14). This mentality was prevalent in other states of recent settlement even though it was never specified in the law. Game wardens in Michigan, Wisconsin, and Minnesota noted later in the century that they often felt it better to ignore violations altogether than to prosecute because a jury of peers would only acquit transgressors. A milestone in the demise of this view may be seen in the case of Joe Knowles, who entered the Maine woods in 1913 for two months, leaving behind all the accoutrements of his civilized life. The event attracted considerable publicity, and, to feed the press, Knowles sent information out of the woods written in charcoal on birchbark. The Boston *Post* gave headlines to a report that he had lured a bear into a pit, killed it with a club, and made a coat from its skin. Upon his emergence from the woods, the Maine Fish and Game Commission fined him $205 for hunting bear out of season [Roderick Nash, *Wilderness and the American Mind* (New Haven: Yale University Press, 1967), pp. 141-42].

8. The general charge to the secretary fails even to mention wildlife. He is directed to "provide for the preservation, from injury or spoilation, of all timber, mineral deposits, natural curiosities, or wonders. . . ." 9. Stat. 17, 32. Cited in John Ise, *Our National Park Policy: A Critical History* (Baltimore: Johns Hopkins University Press, 1961), p. 19.

9. *Fur, Fin, and Feather, A Compilation of Game Laws*, p. 114.

10. Ibid., p. 116.

11. See Chapter 1, text at notes 65-67.

12. On Blooming Grove Park Association, see Chapter 4, text at notes 25-28.

13. See Chapter 2, text at notes 50-55, for examples of rules that were established among market hunters.

14. For a discussion of the growth of due process and police power and their roles in the developing creed of the gospel of wealth, see Ralph Henry Gabriel, *The Course of American Democratic Thought: An Intellectual History Since 1815* (New York: The Ronald Press, 1940), pp. 216-33.

15. As Coggins and Smith observe, there is, in the present day, no "law of wildlife" in the sense that there is property law or tax law [George Cameron Coggins and Deborah Lyndall Smith, "The Emerging Law of Wildlife: A Narrative Bibliography," *Environmental Law*, 6 (1975), 583].

16. Gilmore, following Karl Llewellyn, observes a period of legal formalism spanning the years between the Civil War and World War I, which separates two periods of innovation. During this period, Blackstone's condensation of the common law, written during an earlier formalist period, had great appeal. Although the responsiveness of law to political pressure was denied, the very movements into and out of this period belie this view [Grant Gilmore, *The Ages of American Law* (New Haven: Yale University Press, 1977), passim]. The early innovative period is the subject of Morton Horwitz's comprehensive study, *The Transformation of American Law, 1790-1860* (Cambridge: Mass. Harvard University Press, 1977).

17. 252 U.S. 433. See also Charles A. Lofgren, "*Missouri* v. *Holland* in Historical Perspective," *The Supreme Court Review* (1975), p. 99. On Holmes, see Gilmore, *Ages of American Law*, pp. 48-56.

18. These include *per industrium hominis*, where man takes or confines wild animals, *per impotentiam*, where man possesses immature animals by their inability to escape from his property, and *propter privilegium*, where a landowner had been granted exclusive privilege of property in animals within his domain [John H. Ingham, *The Law of Animals; A Treatise on Property in Animals Wild and Domestic and the Rights and Responsibilities Arising Therefrom* (Philadelphia: T. and J. W. Johnson & Co., 1900), pp. 2-3]. The latter of these qualified properties refers to royal grants and is not applicable in the United States. To these must be added a fourth, *ratione soli*, which, as established in the English case of *Blades* v. *Higgs* (1865), grants to the landowner exclusive rights to capture animals *ferae naturae* on his own land. In the United States, this qualified property exists as against other individuals, but not against the state.

All of these classes of qualified property in wild animals depend to some degree on possession. The implication, then, is that upon escape, the animals again become common property. This assumption might be modified, however, could it be claimed that during the period of possession, the animals in question had ceased to be wild. It is thus important to distinguish between animals *ferae naturae*, which would presumably become common property upon escape, and animals *domitae naturae* (of a domestic nature), the private property in which would remain upon escape. However, by the principle of *animus revertendi*, animals which have a habit of returning of their own volition to the control of an individual, as do bees to a hive, remain his property even though they are at times away from his domain and out of his direct control. This principle was modified in response to the controversy over the ownership of the seals in the Bering Sea. The United States argued that seals which annually migrate to and from United States territory were thereby the sole property of the United States regardless of where they may exist at a particular time. The British argument, which was to become the basis of the ruling, was that *animus revertendi* applied only where animals were induced to return to a location by artificial means, and thus did not apply to naturally migrating animals like seals.

19. This treatment follows Chester Kirby, "The English Game Law System," *American Historical Review* 38 (January 1933), 240-62. See also Austin Lane Poole, *From Domesday Book to Magna Carta, 1087-1216*, 2d ed. (Oxford: The Clarendon Press, 1955), pp. 28-35; and William Blackstone, *Commentaries on the Laws of England, Together with Notes Adapting the Work to the American Student*, by John L. Wendell, 3 vols. (New York: Harper & Brothers, 1847), II, 48-50, 491-501, to which is appended the commentary of Christian.

20. The first law of this kind, passed in 1390, limited to those owning real estate with a rental value of more than forty shillings per year, the killing of hares, conies, or deer and the keeping of dogs. The content of the law changed during the next three centuries, though its general character remained fixed. By 1671, a refined version emerged which would serve as the code for a century and a half. Access to game was limited by this law to four categories of individuals: those with real property worth in excess of one hundred pounds annually; those who were sons or heirs of an esquire or of a person of higher degree; those with leases of ninety-nine years or more, worth in excess of one hundred pounds annually; and the owners of previously existing franchises, granted to their ancestors by a previous king.

21. By 1775, almost as many copies of *Commentaries* had been sold in the American colonies as in England [John C. Miller, *The First Frontier: Life in Colonial America* (New York: Delacorte Press, 1966), p. 263].

22. This discussion follows Connery, *Governmental Problems*, pp. 58-63, and *Forest and Stream* 41 (August 16, 1883), 41-43. See also Thomas A. Lund, "Early American Wildlife Law," *New York University Law Review* 51 (November 1976).

23. As Connery observes, "by the beginning of the sixteenth century the prerogative of afforesting the land of subjects had so far fallen into disuse that when Henry VIII wished to afforest the land around Hampton Court, he obtained statutory authority and provided compensation for the tenants of the land" (*Governmental Problems*, p. 60).

24. An Attorney, *A View of the Principal Parts of the Most Important Statutes Relating to Game: with Explanatory Cases and Observations* (London: J. Ellis, 1801), pp. 5-6.

25. Connery, *Governmental Problems*, p. 63. The case concerned the plaintiff, Blades, a game dealer, who brought action against the defendants for taking from him rabbits that he purchased on consignment. The defendants argued that the rabbits were the property of their master, the marquis of Exeter, because they had been taken from his property.

26. 60 N.Y. 10. This case was successfully prosecuted by Charles Whitehead, counsel for the New York Association for many years, member of the Boone and Crockett Club, and ardent conservationist until his death in 1903. George Bird Grinnell has remarked that "with Royal Phelps and a few others, he originated active game protection in this country, gave it its first impetus, and got a court decision—the Phelps-Racey case—which has formed the basis of almost all legal work done in game protection in the United States." Grinnell's remarks are found in a letter that accompanies Charles Hallock's photo album, now located in the Yale University Library Manuscript Collection. Charles Sheldon, into whose hands the album fell, sent it to Grinnell in 1923 with the hope that he might be able to provide background information on some of the individuals whose portraits were included therein.

Papers relating to this and other cases prosecuted on behalf of the New York Association for the Protection of Game may be found in the Manuscript Collection of the State Library of New York at Albany.

27. 60 N.Y. 14. The state may even forbid the sale of artificially reared trout during periods when the possession or sale of wild trout is prohibited [*Commonwealth* v. *Gilbert*, 160 Mass. 157 (1893)]. The Massachusetts Supreme Court ruled that such a law was justified on

> account of the difficulty in distinguishing between trout which had been artificially propagated or maintained, and other trout. . . .
>
> Such laws are not to be held unreasonable because owners of property may thereby to some extent be restricted in its use. It has often been declared that all property is acquired and held under the tacit condition that it shall not be so used as to destroy or greatly impair the public rights and interests of the community.

See Chapter 6 for the views of marketmen on this and similar restrictions.

28. *Magner* v. *People*, 97 Ill. 334.

29. *State* v. *Rodman*, 58 Minn. 400. See also Connery, *Governmental Problems*, p. 55.

30. 161 U. S. 519. The Court's previous decisions on related matters, *Corfield* v. *Coryell*, 6 Fed. 546 (1823) and *McCready* v. *Virginia*, 94 U. S. 395 (1877), both of which upheld the right of the state to prohibit nonresidents from harvesting oysters in state waters, depended directly on the state's ownership of the land from the low water mark to the three-mile limit, which had passed to the states from the English Crown. For wildlife living on land not owned by the state, this decision would not necessarily apply. (see Connery, *Governmental Problems*, p. 55).

31. Some courts, in the absence of support from the ownership doctrine, could not uphold such exercises of state power. The Kansas Supreme Court, in *State* v. *Saunders*, 19 Kansas 127 (1877), observed that the justices of the U. S. Supreme Court "have been groping their way cautiously, but darkly, in endeavoring to ascertain" the exact meaning of the commerce clause.

> We think however that amidst all their conflicts and wanderings they have finally settled, among other things, that no state can pass a law . . . which will directly interfere with the free transportation, from one state to another, or through a state, of anything which is or may be a subject of inter-state commerce.

The decision of the district court was reversed, and the law prohibiting the export of game was overturned. A similar view is expressed in *Territory* v. *Evans*, 2 Id. 658 (1890).

32. 161 U. S. 529-30. This view finds earlier expression in *American Express Company* v. *People*, 133 Ill. 649 (1890), *State* v. *Geer*, 61 Conn. 144 (1891), and *Organ* v.*State*, 56 Ark. 267 (1892). For the view that Justice White, in this and other decisions—for example, *Ward* v. *Race Horse*, 163 U. S. 504 (1896) and *The Abby Dodge*, 223 U. S. 166 (1912)—represents a short term and somewhat isolated judicial stance in his support of the state ownership doctrine. See Council on Environmental Quality, *The Evolution of National Wildlife Law*, 1977, pp. 18-23, 54-56.

33. *State* v. *Randolph*, 1 Mo. App. 17 (1876) and *Roth* v. *State*, 51 Ohio 212 (1894).

34. This was not suggested as a basis for interference with the authority of the state until the federal government sought control over migratory birds in the twentieth century.

35. *Commonwealth* v. *Wilkinson*, 139 Pa. 298.

36. "An Act to enlarge the powers of the Department of Agriculture, prohibit the transportation by interstate commerce of game killed in violation of local laws, and for other purposes" (31 Stat. 187). Note the similarity in wording to the Wilson Original Package Act of 1890, designed to prohibit the interstate passage of alcoholic beverages. (Council on Environmental Quality, *The Evolution of National Wildlife Law*, 1977, p. 112, n. 13). On the passage of the Lacey Act, see Chapter 6, text at notes 188-200.

37. The laws of most states did not distinguish among the many species of duck. See James Tober, "The Allocation of Wildlife Resources in the United States, 1850-1900" (Ph.D. diss., Yale University, 1973), pp. 291-303, for comparable charts showing the growth of seasonal protection for the woodcock, prairie chicken (pinnated grouse), quail, and ruffed grouse.

38. In twenty-two states having closed seasons for deer in 1871 (not including the closed term in Massachusetts), there were seventeen different seasons. There were eight different beginning dates, from December 1 to March 1, and seven different ending dates, from July 1 to October 1. While it is not to be expected that states with widely divergent climates would have identical seasons, it should be noted that the seasons vary almost as much within as across regions. The closed season in New Hampshire, for example, was two months

shorter than that in Maine, and the closed season in Nebraska was three months longer than that in Kansas. By 1882, the uniformity had not improved. According to *Forest and Stream's* table of open seasons, [18 (July 20, 1882), 488-89], of the thirty-six states (and Washington, D.C.) with closed seasons for deer (excluding Vermont and New Jersey, which had closed terms until November 1886 and October 1884, respectively), there were twenty-six different closed seasons, beginning between October 1 and March 1 and ending between April 1 and October 20. The increased spread of seasons was due in part to the increased geographical range of states providing protection. By 1901, the number of states with closed terms had increased to eleven. Of the thirty-five states (and Washington, D.C.) with closed seasons for deer in that year, there were twenty-eight different periods, beginning between October 1 and March 15 and ending between August 1 and November 15.

39. See Chapter 6, text at notes 22-34, for a chronology of these meetings.

40. An 1855 Massachusetts statute (Ch. 197, pp. 615-16) provided that the game law shall not extend to

> any town in which the inhabitants shall, at their annual meeting in any year, vote to suspend the operations thereof in whole or in part, and for such terms of time, not exceeding one year, as they shall think expedient.

In 1869 (Ch. 246, pp. 581-83), this option was removed and replaced by a section which made it encumbent upon local officials to enforce the law within their domains.

41. A correspondent to *Forest and Stream* offered this scenario:

> A Vermonter, say Sam Lovel of Danvis, a stranger in Maryland, is caught shooting partridges. . . . in Dorchester County without first having taken out his $5 license; he is arrested, hauled before the local justice of the peace and convicted and fined $25. Not having the money to pay his fine he is about to be led away to the lock-up for ten days' imprisonment, when he recognizes in the signature of the justice affixed to the commitment paper the name of a Marylander who is a cousin of his Danvis deceased wife's sister's husband; and forthwith the Yankee claims his liberty, asserting himself to be "a connection by marriage of a

bona fide citizen of said county," and as such, according to the letter of the statute, exempt from the non-resident shooting law, its pains and penalties. The facts of the connection by marriage being proven to the satisfaction of the magistrate, the prisoner is not only discharged, but with true Maryland hospitality is invited to accompany the justice home and puts in a week of partridge shooting... [40 (April 6, 1893), 291).

42. U.S. Department of Agriculture, Division of Biological Survey, Bulletin 16, *Digest of Game Laws for 1901*, by T. S. Palmer and H. W. Olds (Washington, D.C.: G.P.O., 1901), Table II.

43. The following data are generally derived from *Digest of Game Laws for 1901*.

44. Some particular prohibitions are of interest. Idaho and Wyoming, for example, specifically excluded mounted heads and stuffed specimens from the export prohibition but did not restrict the preparation of captured game within the state. In addition to supporting local taxidermy, the law encouraged trophy hunting relative to meat hunting and therefore favored the capture of males. Iowa permitted a nonresident to remove from the state up to twenty-five game animals or birds if he accompanied them in transit. Oregon allowed residents of Washington to take home one day's bag. Sixteen states, some of which required permit or license, made exceptions for the export of game shipped alive for propagation.

45. The most obvious explanation for the prior existence of fish commissions is that the artificial propagation of fish was demonstrated relatively early. In 1857, Massachusetts formed the Commission on Artificial Propagation of Fish, which reported general evidence on the breeding of fish and reproduced an important article on methodology by the French fish culturist Jules Haime [Massachusetts, *Report of the Commissioners Appointed under Resolve of 1856, Chap. 58, concerning the Artificial Propagation of Fish* (Boston, 1857)]. The commission noted: "We believe there are many farms on the hilly and mountainous parts of Massachusetts, containing trout streams, that, with a little pains might be made to yield a greater income in this way, than the land itself" (ibid., p. 9). Compare this to the statement of the Massachusetts Fish Commission thirty years later: "Ground game once exhausted cannot be replaced by artificial propagation as the fishes can" [Massachusetts, *Report of the Commissioners on Inland Fisheries and Game for 1887* (Boston, 1887), p. 32].

The New England states were particularly concerned over the condition of the Connecticut River fishery. It was clear that since the river, as well as the fish of the river, passed through four states, overfishing, pollution, and dams within the borders of any one of them threatened the fishery for the others. In 1866, representatives of Massachusetts, Connecticut, New Hampshire, and Vermont met and agreed that the two northern states would stock the river; Massachusetts would provide fishways over all dams along the river within its borders; and Connecticut would prohibit all pounds at the mouth of the river. As with similar voluntary arrangements among sportsmen's associations, the effort failed. It led, nonetheless, to the establishment of separate fish commissions in the New England states. Lund reports an earlier attempt by Massachusetts to seek the cooperation of New Hampshire in the protection of fisheries in 1783 and again in 1790 ("Early American Wildlife," n. 144).

Concern for the dimunition of food fishes in U.S. waters led, in 1871, to the creation of the United States Commission of Fish and Fisheries. Under the direction of Spencer Baird, who held the office of commissioner through 1887, the agency engaged in substantial work on fish culture. The appropriation for this activity increased from $30,000 in 1876 to $161,000 in 1887 [Robert H. Connery, *Governmental Problems in Wild Life Conservation* (New York: Columbia University Press, 1935), chapter 6]. Among the practical results of this work was the distribution of food fishes across the country. The existence of the federal commission and the consequent availability of fish eggs and fry, as well as of scientific information on fisheries, were no doubt largely responsible for the spate of state agencies created in the 1870s. See also John F. Reiger, *American Sportsmen and the Origins of Conservation* (New York: Winchester Press, 1975), pp. 52-56.

46. For a detailed treatment of federal wildlife law, see Council on Environmental Quality, *The Evolution of National Wildlife Law*.

47. 58th Congress, H. R. 15601, quoted in James B. Trefethen, *Crusade for Wildlife: Highlights in Conservation Progress* (Harrisburg and New York: The Stackpole Company and the Boone and Crockett Club, 1961), p. 166.

48. More detailed accounts of these and subsequent events may be found in Trefethen, *Crusade for Wildlife*; John C. Phillips, *Migratory Bird Protection in North America: The History of Control by the United States Federal Government and a Sketch of the Treaty with Great Britain.* Special Publication of the American Committee for International

Wildlife Protection, Vol. I, No. 4 (1934); American Game Protective and Propagation Association, *The Game of a Continent Ours to Protect* (1913?); and Thomas Gilbert Pearson, *Adventures in Bird Protection* (New York: D. Appleton-Century Company, 1937).

49. *United States* v. *McCullagh*, 221 F. 288 (D. Kan. 1915) and *United States* v. *Shauver*, 214 F. 154 (E. D. Ark. 1914).

50. William S. Boyd, "Federal Protection of Endangered Species," *Stanford Law Review*. 22 (June 1970), 1294.

51. 252 U. S. 434 (1920). For a retrospective look at *Missouri v. Holland*, with special emphasis on its implications for federal treaty power, see Charles A. Lofgren, "*Missouri v. Holland* in Historical Perspective," *1975-Supreme Court Review*, ed. Philip R. Kurland (Chicago: University of Chicago Press, 1976), pp. 77-122.

52. *Toomer* v. *Witsell* 334 U. S. 384; *Takahashi* v. *Fish and Game Commission* 334 U.S. 410. Sufficient grounds for decision might have been found in the "equal protection" clause of the Fourteenth Amendment. These cases generated considerable commentary. See, for example, Nicholas V. Olds and Harold W. Glassen, "Do States Still Own Their Game and Fish?" *Michigan State Bar Journal* 30 (1951), 16-23; H. M. Durham and R. B. Herrington, "State Ownership of Fish and Game," *Georgetown Law Journal* 38 (1950), 652-58; Eugene F. Kobey, "Discrimination by State Against Non-Residents' Hunting and Fishing Privileges," *Marquette Law Review* 33 (1950), 192-95; and Irvin B. Charne, "Fish and Game—Power of State to Regulate Taking Of," *Wisconsin Law Review* (January 1949), 181-84.

53. Olds and Glassen, "Do States Still Own Their Game and Fish?" p. 22.

54. 47 U.S.L.W. 4447; 9 *ELR* 10106.

55. This section follows Boyd, "Federal Protection of Endangered Species," pp. 1295-1302. See also Sandra Jo Craig, "Wildlife in the National Parks," *Natural Resources Journal* 12 (October 1972), 627-32; James R. Dickens, "The Law and Endangered Species of Wildlife," *Gonzaga Law Review* 9 (Fall 1973), 57-115; George Cameron Coggins, "Legal Protection for Marine Mammals; An Overview of Innovative Resource Conservation Legislation," *Environmental Law* 6 (1975) 1-59; "Federal Preemption: A New Method for Invalidating State Laws Designed to Protect Endangered Species," *University of Colorado Law Review* 47 (1976), 261-78; Glen O. Robinson, *The Forest Service*, Chapter 8, "Wildlife" (Baltimore: Johns Hopkins University Press, 1975); George Cameron Coggins and William H. Hensley, "Constitutional

Limits on Federal Power to Protect and Manage Wildlife: Is the Endangered Species Act Endangered?" *Iowa Law Review* 61 (June 1976),
1099-1152; Council on Environmental Quality, *Evolution of National
Wildlife Law*, 1977.

56. For example, *Bailey* v. *Holland*, 126 F 2nd 317 (4th Cir. 1942),
specifically provided that Congress might protect birds adjacent to a
federal wildlife refuge as a means of protecting them as they entered
and left public land.

57. Boyd, "Federal Protection of Endangered Species," p. 1299.

58. Ibid.

59. Since some individuals of virtually every species cross state
lines, this would provide extensive authority.

60. Boyd, "Federal Protection of Endangered Species," p. 1301.

61. Coggins and Hensley offer an additional, implied, basis for
federal control over wildlife in the protection of species with symbolic
value to the nation. This authority, never argued in court, appears
important in the Bald Eagle Protection Act, passed in 1940 and
amended in 1962 to include golden eagles, and in the Wild and Free-
Roaming Horses and Burros Act of 1971 ("Constitutional Limits," pp.
1139-43). The latter act has recently survived a Supreme Court test
[*Kleppe* v. *New Mexico*, 44 U. S. L. W. 4878 (1976)]. The Court unanimously held the act a constitutional exercise of congressional power
under the property clause (ibid., editor's note, p. 1152).

62. Two early cases generally cited as upholding nonresident discriminations under the doctrine of state ownership do not deal
squarely with the issue. In *Allen* v. *Wyckoff*, 48 N. J. L. 90 (1886), the
court upheld the provision contained in the bylaws of a state-
chartered association for the protection of game that nonresidents
must join the association having jurisdiction over the desired hunting
ground in order to legally hunt there. The court refused to confront
the possible violation of the privileges and immunities clause of the
Constitution by noting the absence of evidence that the defendant
was a citizen of any particular state or even of the United States. In *In
Re Eberle*, 98 Fed. 295 (1899), an Iowa resident was part of a corporation
which owned the Crystal Lake Club in Illinois. He was arrested and
fined, on the land of the club, for failing to comply with the Illinois
nonresident license law. The Court upheld the conviction, not on
general grounds but on the basis that a corporation, not Eberle,
owned the land. In 1914, the U.S. Supreme Court, in *Patsone* v.
Pennsylvania, 232 U. S. 138, upheld the right of the state to discrim-

inate against nonresidents and further established its right to prohibit aliens not only from hunting within its boundaries but even from owning guns used in hunting.

It appears, however, that even with the weakened state ownership doctrine, the state may, through exercise of its police power, discriminate between residents and nonresidents if such discrimination is neither arbitrary nor unreasonable and if it "tends to accomplish or has a substantial connection with a valid state purpose" [Dan Riggs, "Constitutional Law-Wyoming's Guide Law: Non-resident Hunters on Public Lands and Collateral Issues," *Land and Water Law Review* 9 (1974), 169-83]. See also *Baldwin* v. *Fish and Game Commission*, 436 U.S. 371 (1978), in which the Supreme Court upheld Montana's differential license fee structure on the ground that recreational hunting was not a fundamental right protected by the privileges and immunities clause (9 *ELR* 10107).

6

PUBLIC POLICY FOR WILDLIFE

The trouble in formulating game laws seems to be that parties interesting themselves in the matter want only such restrictions placed on the killing of game as will restrict the other fellow leaving their own privileges intact.
-letter to *Western Field and Stream*, 1897

The gradual but steady success of sportsmen in achieving a realignment of property rights in wildlife ultimately depended on their high degree of organization and their ready access to the political process, but the receptivity of legislatures and courts grew out of a broader social change that sportsmen and their allies were able to exploit. The disappearance of the frontier symbolized a shift in popular ideology from extraction toward preservation. It was exemplified by the great interest in song birds and the consequent campaign against the plume hunters. The nineteenth-century growth in scientific knowledge contributed to a shift in favor of wildlife on the farm and foreshadowed the early-twentieth-century conservation movement that identified the end of an era of inexhaustibility by supporting the wise use calculations of experts and by recognizing public values in resources that had always been considered only as potential private goods.

Avenues available for the expression of sportsmen's preferences were limited and had been laid out over two centuries of settlement and growth. The reservation of game animals for a sporting class was early ruled out of consideration by constitutional guarantees of open access to the hunt and by an egalitarian world view. The Jeffersonian republic, made inevitable by a great abundance of land and a deliberate avoidance of European privilege, created a nation of freeholders, the final manifestation of which is the suburban single-family home and the vacation cottage. The transformation of the landscape, both public and private, was virtually unconstrained throughout the nineteenth century. On private parcels, the imposition of land-use controls in the name of wildlife management would have severely restricted private income in exchange for obscure benefits that would have disproportionately favored sportsmen. The clarity and breadth of private property rights in land represented the unwillingness of the nation to make the exchange and, perhaps as well, the inability to enforce it. The market offered some reconciliation in those cases where mutual gains were perceived, but the extent of wildlife habitat entirely beyond the control of sporting interests was tremendous. On public lands, sportsmen were offered a clearer access to management decisions through normal political channels, but, even here, the lack of information, conflicts with management for other purposes, and a poorly developed administrative and enforcement structure precluded all but the barest recognition of the needs of wildlife on the land.

Instead, the proprietary interest of the states in wildlife and their well-established tradition of intervention in the time and place of the hunt logically directed the controversy into this regulatory arena. Sportsmen and their allies sought increased control over wildlife through the broadening of existing restrictions, through the development of new types of restrictions which, while only marginally pushing out the regulatory frontier, might significantly affect animal populations and extractive behavior, through the creation of mechanisms to enforce existing regulations, and, finally, through a broad campaign designed to increase the public acceptance of the propriety of these changes.

This chapter examines the foundations of a public policy for wildlife as it was articulated and promoted by a small political elite with the selective support of broad constituencies and against the background of ideology, resource abundance and decline, and institutions detailed above.

THE LEGISLATIVE CONTEXT

The most consistent force behind the passage of game legislation in the nineteenth century was exerted by sportsmen. Their direct influence varied from state to state and region to region. In New York, game legislation was often drafted directly by the New York (City) Association for the Protection of Game. An example is the important 1871 game law codification, the "great object" of which was "to bring the markets of the City under a reasonable control for the Preservation of Game." Royal Phelps, Robert B. Roosevelt, and Charles Whitehead, on behalf of the association, wrote to C. W. Hutchinson of Utica, president of the State Sportsman's Association. They indicated that they had arranged for the introduction of the bill on the first day of the session and expressed the hope that they would "find a hearty support from the members of the Legislature from the Rural Districts, and at the same time a dispostion to waive unimportant local interests." They urged that the matter be managed quietly and kept out of the papers "in order to avoid factuous opposition at Albany from the 'Market Men' of this City."[1]

But in spite of successes of this kind, the *American Sportsman* was certain that the general influence of sportsmen was limited by the "fact that few sportsmen are legislators or few legislators sportsmen."[2] If it were simply that nonsporting legislators were indifferent to the concerns of sportsmen, adequate education might suffice, especially if it was augmented by new fish and game committees, the membership of which would heavily represent the few sportsmen in the legislature. But it was not simple indifference. Legislators who were not "sportsmen" were hunters nonetheless who might have been accustomed to behavior discouraged by sportsmen. Nor were the fish and game committees necessarily successful; they only served to identify the individuals to whom competing interests ought to direct their influence.

Thus, far from being in the hands of sportsmen, legislatures were influenced by a variety of constituencies on the questions of the game laws. An 1881 protective bill supported by the Massachusetts Fish and Game Protective Society was defeated by a large margin because "game legislation is too much in the hands of know-nothings, and know-nothings are in the hands of a class of men who are concerned not to preserve the game, but to squeeze the almighty dollar out of it as it goes."[3] A decade later the sportsmen's interests in Massachusetts were opposed by the "wholesale marketmen, the cold storage men, the fertilizer men, the oil men, the net and twine organization, [and] the United States Menhaden Oil and Guano Association." Sportsmen may be men of great personal influence, but they were only "guerillas opposed to regular troopers."[4]

Whatever influence sportsmen actually had, many felt they should be accorded more. *Science* magazine's 1886 supplement on the plight of birds in the United States concluded that the

game-birds should be left to the care of sportsmen and game-protective associations, since self-interest on the part of the more intelligent sportsmen will dictate more or less wise legislation for the preservation of the birds on which their sport depends.[5]

Not unexpectedly, sporting journals had already articulated this view. Citing "the highest English Authority," *Forest and Stream*, in 1874, related "the known fact that all the best measures for the protection of game . . . must always emanate from those who shoot and fish for their pleasure."[6] Twenty years later, the journal observed that those

who seek to secure game legislation opposed to the sportsman's interest may possibly be sincere in believing that they are working for the public good; but they are not. . . . Game laws can benefit the community only as, and in such degree as, they are in the interest of sportsmen.[7]

It was not long before the influence of sportsmen was under attack from a new direction. Just as sportsmen accused market hunters of depriving them of the opportunity to partake in the "highest" form of the hunt by irresponsible killing for the market, William Hornaday was "shocked by the accumulation of evidence showing that all over our country and Canada fully nine-tenths of our protective laws have practically been dictated by the killers of game." The well-appointed and influential sporting gentleman has "no more 'right' to kill a covey of quail on Long Island than my milkman has to elect that it shall be let alone for the pleasure of his children."[8] The very constituencies that sportsmen had rallied in their battle for control over wildlife policy were now expressing demands that conflicted with their own. But the sportsman's position was by that time well entrenched, and, even now, over half a century later, sporting interests dominate management except in areas specifically designated as sanctuaries.[9]

Hornaday's views, of course, reflect his own preferences, but they are an expression too of the wise-use doctrine of the conservation movement, which promoted rational planning in the use of natural resources.[10] But prior to the 1930s, there was little information that might have provided the basis for the reasoned judgment that Hornaday apparently sought.[11] Whatever was known, was known by sportsmen as well as by anyone else. Others who might have assumed the role of "experts" during the nineteenth century were the naturalists and the state fish and game officials. The naturalists, for the most part, were not wholly distinguishable from the sportsmen in their perspective on the problems of wildlife. Exceptions were John Muir and John Burroughs, who generally made their influence felt in matters of wilderness preservation rather than in matters of game protection. The fish and game commissioners and agents most often had only small influence on legislation. They had no formal place in the legislative process, but, beyond this, they often had no prior interest in or knowledge of wild animals and were, in many states, political appointees. Commissioners disagreed over their appropriate legislative roles. The Connecticut and New York commissioners agreed that they ought to have a voice in the formulation of the game laws, but the Michigan warden felt that it was not his "province to advise what the seasons on various kinds of game should be."[12]

Conflict among the interest groups seeking preferred access to wild, and particularly game, animals did not lead to a stalemate in which forces matched one another at every turn, frustrating the emergence of legislation acceptable to any group. Rather, the volume of legislation was large and generally aimed in the direction of reducing the hunt through limiting seasons, technologies, bags, and conditions of sale and transport. But while the path was identifiable, it was hardly direct. There were zig-zags and backtracks. The frequent change in the law was due in part to a learning process in which groups with particular goals would experiment with regulations, observe the consequences, and correct constraints accordingly. Unfortunately, the learning was minimal because constraints may either have reinforced or countered natural wildlife population cycles. Perhaps the frequent change in the law was more the result of the alternate support that the legislature gave to competing interests. In some states, this pattern produced nearly annual changes in the hunting seasons. Desired reform did not even depend on mounting a state-wide campaign. Interest groups with a regional or local power base might simply work to have enacted special legislation to except a particular county or a single body of water from existing law.

Because New York residents had a wide diversity of interests and the land was geographically diverse, the state was particularly susceptible to frequent change in the law. *Forest and Stream* pleaded with the legislature to leave the law alone "long enough for the people of the State to find out what it is."[13] The journal later noted that "the taxpayers of New York pay on average $734 for each new statute enacted by the Legislature. Many a local fish and game law gained at this price is actually worth less than seven dollars and thirty-four cents."[14] During the 1880s and 1890s especially, the number of fish and game bills introduced in the New York legislature was easily large enough to justify such concern. *Forest and Stream* counted fifty-one in 1887, fifty-seven in 1888, forty-eight in 1889, and eighty-eight in 1897.[15] Although most of them never found their way to the governor's desk, the number that did and were signed averaged seven per year in the 1870s, eight per year in the 1880s, and seventeen per year in the 1890s.[16]

The difficulty in developing even a semi-permanent code of game laws within a state was magnified by the difficulty in coordinating the development of game laws among states. Beginning in the 1870s, the interstate uniformity of seasons, important both in simplifying the law and in reducing the mobility of hunters and killed game, was a topic of great concern among sportsmen.[17] *Forest and Stream*, in noting the disparities existing in 1879, put forth a "Simplified Plan for Uniformity of Close Seasons—Legislation Made Easy." Simple and sane management required a single closed season for each of a number of important game species. The journal could find only a marginal relationship between existing seasons and climate and saw, therefore, no reason not to take the most commonly occurring dates as those to which all states should subscribe.

We must consider that any exception in favor of any locality or kind of game leaves a wide loophole of escape for any man who wishes to evade the laws. Make the close season uniform throughout the country and let it apply to all kinds of game, and the man who is found abroad in the interval with dog, gun, and hunting paraphernalia, will have a hard job to acquit himself of deliberate intent to break the law.[18]

Throughout the remainder of the century, more or less frequent pleas were made for the establishment of uniform state laws concerning hunting seasons. The most notable of these was made in 1897, again by Charles Hallock, then editor of the *Western Field and Stream*. The plan, which came to be known as the "Hallock Code," divided the country into three regions, each with its own uniform laws.[19] The journal boldly likened the game laws to the Ten Commandments, noting that the Commandments would not have survived were they filled with exceptions for localities and individuals.[20] *The New York Times* blamed the entire game shortage on the failure of the nation to heed pleas for uniform laws in the 1870s when Hallock had first put forth the importance of the concept. Had this first effort crystallized, "the buffalo would have been saved, the wild pigeon would have been saved, and the whole game category been cared for."[21]

It was for the purpose of providing a forum for the spread and adoption of uniformity in game legislation that William C. Barrett, Charles H. Whitehead, and Charles Hallock, representing the New York Association for the Protection of Game, called for a national convention of sportsmen to be held late in 1874. The New York State Sportsmen's Association had pre-empted the city organization's call, however, and planned a national meeting for September 1874, to be held in Niagara, New York. The city organization felt obligated to defer as a subordinate club even though Hallock "did not think the Niagara convention was composed of the very best materials in the country."[22]

The national convention was successful neither in establishing uniform legislation nor in building an organization that might have future success. The tone of the meeting is indicated by its early action in voting down the name National Association for the Protection of Fish and Game in favor of the National Sportsmen's Association. Though notices of the meeting were sent to every domestic newspaper and all known sporting clubs, only eight-six delegates from sixteen states and the District of Columbia attended. Of this number, thirty-two represented New York clubs and thirteen represented Ohio clubs. *The New York Times* suggested that the whole affair was called by arms manufacturers in order to stage a grand pigeon shoot. The *American Sportsman*, in a dubious refutation, called attention to the fact that in attendance were two editors, four lawyers, two bankers, a coach proprietor, a legislator, a dealer in coal, a merchant, a clergyman, and three men of independent means. The journal noted further a singular likeness between the *Times* and a "sporting contemporary in their support for the New York City association."[23]

It appears that good intentions were abundant, however. The association adopted a resolution seeking

first, the protection of game and fish in all the States and Territories, by procuring the passage in each State, and in the Congress of the United States, of uniform co-operation and consistent laws, . . . second, through subordinate organizations, to insure a rigid enforcement of all the game and fish protection laws now or hereafter to be enacted; third, to secure by and through proper legislation the right of property in useful hunting dogs, making them, when stolen, the subject of larceny, or when wantonly killed or maimed the subject of a misdemeanor; forth, to secure, through and by proper legislation, the passage of laws prohibiting at any time and season the killing or destruction of all song and non-game birds; to organize and consolidate, under our State jurisdiction, game and fish protective clubs or associations in each State and Territory, to act under the jurisdiction of the National Association, in securing and enforcing proper protective legislation.[24]

The following June, the second annual convention was held, at which the sportsmen of twelve states were represented by forty-one delegates.[25] At subsequent meetings, attendance continued to decline. In September 1876, sportsmen of nine states were represented, and the following June in Syracuse only six delegates appeared. Nine short of a quorum, the convention conducted no business. The meeting of the following year barely attracted notice in the columns of *Forest and Stream*.[26]

Interest in a multistate organization of sportsmen surfaced from time to time through the turn of the century.[27] Early in 1884, representatives of the fish and game protective associations of the New England states met for the purpose of drafting uniform regional legislation. The delegates agreed to uniform closed seasons for the important game species, to full protection for non-game birds (except birds of prey, sparrows, crow, blackbird, and jay), and the prohibitions of trapping, the use of ferrets, jack lights, and swivel guns. Finally, the commissioners of fisheries were to be granted authority over game animals.[28] Although some of these proposals were incorporated into the laws of individual states, no greater uniformity was achieved as the result of this effort.

In 1885, a group of St. Louis sportsmen called for the organization of the National Association for the Protection of Game, Birds and Fish. Members were invited not only from among the sportsmen of all states but from among fish commissioners holding state office and game dealers representing large, urban markets.[29] The first meeting took place in the fall of the year and was attended by sportsmen from several Midwestern states as well as by several fish commissioners and game dealers. The membership called for uniform seasons east of the Rockies similar to those proposed by the New England commissioners the previous year, for a bag limit on all species of protected birds of twenty-five per day, and for the appointment of a number of state-level game protectors, each delegation of which was to be supervised by a chief warden.[30] These suggestions were largely disregarded. The organization persisted through the close of the decade but achieved no notable results.

In reviewing the events of the 1880s, *Forest and Stream* remarked on the "absolute futility of accomplishing anything substantial by convening in national assembly and passing resolutions."[31] When the idea of a national assembly of sportsmen surfaced again in 1893, the journal, thinking back, could find as the only benefit of the previous effort at St. Louis the compilation of a list of sportsmen throughout the country—a list that the promoters subsequently sold to advertisers. *Forest and Stream* therefore saw little gain in repeating the gesture.[32] Nonetheless, a meeting for the purpose of organizing the National Game, Bird and Fish Protective Association was held in Chicago in the fall of 1893. This was attended by only six delegates, four of whom were from that city. "The meeting was a convention only in name, and was national only in the imagination of the individual who originated the scheme and has been its chief promoter. . . . Nothing further may be expected from the impracticable and futile scheme."[33] The first full meeting of the new association was held in Chicago in early 1894, and it continued to meet regularly for several years. Its chief contribution was the role it envisioned for Congress in regulating the interstate shipment of wildlife and in promoting the propagation of game to parallel the work in fish culture undertaken by the U.S. Fish Commission.[34]

Although these early national associations of local sportsmen's groups were notably unsuccessful in achieving their legislative goals, several national associations of individual sportsmen arose toward the end of the century that had a more significant impact on legislation. Their success is explained by a unified administration that could identify policy positions and lobby for change without soliciting the prior support of state or local affiliates and by the social and political stature of the membership.

The most influential of these organizations by far was the Boone and Crockett Club, founded in 1887 at a dinner given by Theodore Roosevelt for a select group of his sporting friends and relatives, including his brother, Elliott, his cousin J. West, and *Forest and Stream's* editor, George Bird Grinnell.[35] The club could count among its early members prominent statesmen, scientists and artists, including Henry Cabot Lodge, Carl Shurz, Owen Wister, Francis Parkman, T. S. Van Dyke, Senator George G. Vest, Albert Bierstadt, D. G. Elliott, Col. Richard Irving Dodge, Gifford Pinchot, J. Pierpont Morgan, and Elihu Root.[36]

Roosevelt explained that the initial purpose of the organization was to protect big game in little-settled regions of the nation, but by the time the formal constitution was drawn up, the objectives had broadened to five:

(1) To promote manly sport with the rifle. (2) To promote travel and exploration in the wild and unknown, or but partially known, por-

tions of the country. (3) To work for the preservation of the large game of this country, and, so far as possible, to further legislation for that purpose, and to assist in enforcing the existing laws. (4) To promote inquiry into, and record observations on the habits and natural history of, the various wild animals. (5) To bring about among the members the interchange of opinions and ideas on hunting, travel, and exploration; on the various kinds of hunting rifles; on the haunts of game animals, etc.

The constitution provided further that "No one shall be eligible for membership who shall not have killed with the rifle in fair chase, by still-hunting or otherwise, at least one individual of one of the various kinds of American large game." The term "fair chase" did not include "killing bear, wolf, or cougar in traps, nor 'fire-hunting,' nor 'crusting' moose, elk, or deer in deep snow, nor killing game from a boat while it is swimming in the water."[37] Roosevelt wrote that the

club is emphatically an association of men who believe that the hardier and manlier the sport is the more attractive it is, and who do not think that there is any place in the ranks of true sportsmen either for the game-butcher, on the one hand, or, on the other, for the man who wishes to do all his shooting in preserves, and to shirk rough hard work.[38]

The club worked to outlaw hunting techniques that it discouraged among its own members, and it was especially active in the battle against hounding deer in New York's Adirondack Mountains. The club interested itself as well in halting the destruction of big game in Yellowstone Park and, with the aid of *Forest and Stream*, was able to secure the protection of the park's wildlife by act of Congress, and to defeat proposals that would have opened regions of the park to private enterprise.[39]

The Boone and Crockett Club exercised great influence over a long period of time primarily by using the individual contacts and personal alliances that its well-placed members were able to make. A second organization, the League of American Sportsmen, was influential during its short life because it effectively mobilized its large membership. The league, founded in 1898, was proposed by George O. Shield's monthly, *Recreation*.[40] Modeled after the League of American Wheelmen, the league established a network of state and county wardens to enforce the game laws and to provide a base for the organization of sportsmen. The proposal drew an enthusiastic response.[41] The formal organizational meeting was held in New York in January 1898, and by December, the league had enrolled 1,051 members who each paid one dollar in dues.[42] By February 1900, divisions had been established in twenty states, and eighty local wardens had been appointed.[43] The club played a major role in the defeat of the section of the New York game law of 1895 that had opened the market to the unlimited sale of game and in the passage of the Lacey Act.[44]

REGULATION AND REFORM

The regulatory forms available in the management of wildlife are limited by the character of the resource itself, by the jurisdiction that may exercise control, and by existing patterns of regulation. Here we consider three forms that occupied the attention of sportsmen, game dealers, landowners, legislatures, and courts throughout the final third of the nineteenth century and that represent the variety of regulatory options at the state level: the technology of the hunt, and, in particular, the hounding of deer; the sale of game and the propriety of its prohibition; and the introduction of hunting licenses as a means to control the hunt and, ultimately, to raise funds for game law enforcement and habitat improvement.

The Limitation of Hunting Techniques

Many of the hunting methods opposed by sportsmen were opposed because they were efficient techniques for market hunters. When the legislative battle was merely between sportsmen and market hunters, sportsmen were generally victorious. By 1901, twenty-seven states had prohibited the use of big guns and/or swivel or punt guns, sixteen states had prohibited the use of some kinds of boats in hunting waterfowl or marsh birds, twenty-two states had prohibited the use of artificial light in hunting big game and/or wild fowl, and twenty-three had prohibited the night-hunting of certain species.[45] When an opposed method was one traditionally employed by rural populations, opposition by sportsmen was seen as the encroachment of city wealth on country lifestyles. Thus, sportsmen often had difficulty in securing and maintaining prohibitions on snaring upland game birds. In 1887, the Massachusetts legislature restored the right of landowners to trap grouse and rabbits on their own property from October 1 to January 1. "The farmers were not going to stand the greediness of sportsmen any longer."[46] But, it was clear to *Forest and Stream*, in spite of the senate "display of codfish eloquence about the constitutional privileges enjoyed by the farmers of Old Massachusetts from the earliest Colonial times," that the real gainers would be the professional market hunters.[47] The Massachusetts commissioners on inland fisheries and game were sympathetic.

It is true that the snare and the trap bring more game to market with less effort than does legitimate and manly pursuit. Like usurious interest, the snare and trap work night and day, constantly and industriously eating away the principle, destroying the stock.[48]

A long and hard-fought battle, and one that well illustrates the articulation of wildlife interests, emerged over the practice of hounding deer in New York. In hounding, dogs are used to drive deer past the waiting hunter, or, alternatively, into the water, whereupon the hunter and his guide can kill the animal with ease from their boat. It was the distaste for this latter form, variously called "water killing" and "water butchery," that generated a large part of the anti-hounding sentiment. Charles Hallock described the practice in the Adirondacks:

[T]he deer is driven until it takes to the water, and when so far from the shore that it cannot return, the hunters row after it, and having approached within a few feet, one of them blows out its brains. When the deer are thin they sink immediately after being shot, and *it is customary for the guide or one of the hunters, if there be two in the boat, to hold the struggling brute by the tail while the other shoots it,* thus saving the carcass. Comment is unnecessary.[49]

Until 1885, the practice was both legal and common in the state.[50] *Forest and Stream*, after years of dodging on its stand, came out firmly against the practice in 1885 with the publication of a petition which had "already been largely signed," requesting the absolute prohibition of chasing or running deer with dogs. The New York State Senate subsequently passed, by a vote of nineteen to two, the anti-hounding bill of General Curtis, an action for which *Forest and Stream* took the majority of credit.[51] The journal had already isolated three groups that it believed might rally for the defeat of the bill: resort owners who thought that prohibition would detract from their business, some city sportsmen who were accustomed to such activity and would view its prohibition as a reduction in their freedom to engage in sport, and some Adirondack residents who hounded deer for the market.[52] The state assembly passed the bill, however, and after some dickering with the governor on matters relating to the giving of evidence in trial, it became law in June 1885. But the small cadre of state agents was ineffective in the face of widespread violation, and protectors were reluctant to prosecute when they knew that local juries would acquit defendants. "J.T.D." wrote to *The New York Times* on the uselessness of legislation "which does not appeal to the common sense of the majority of the resident population."[53] In one case, however, local residents were sufficiently in favor of the law to contribute $245 to pay the salary of a private detective in their region of Herkimer County.[54]

The reality of the prohibition called forth a new unity among the pro-hounding forces. *Forest and Stream* foresaw the conflict and warned that

a movement is on foot to secure the repeal of the non-hounding law. Unless the friends of the deer rally to support the present statute it is

quite probable that a change will be made.... Speaker Husted
. . . is out and out an advocate of deer hounding. He has made up the
game law committees in a way favorable to the hounders.

The question was not one on which reasonable men might
disagree, but one which separated "all intelligent and unselfish Adirondack residents and visitors" from a "small class" of
less respectable men who favored "the disastrous and brutal
waterbutchery."[55]

Several bills were introduced in the 1886 session of the
legislature by representatives of the small class. Although the
bills were generally more restrictive than existing law, they did
provide for hounding during certain periods. The package that
emerged from the state assembly, by a vote of ninety-three to
twenty-three, permitted hounding from September 1 to October 15 during a hunting season that extended from August 15
to November 15.[56] *Forest and Stream* attributed the state assembly's action to political machinations and logrolling. In defer-
rence to the integrity of the legislators, the journal preferred
this interpretation to the one that had them being influenced
by the "ridiculous speech" of Floyd J. Hadley, the chairman of
the game committee in the assembly, the thrust of which was
that the prohibition of hounding had caused a decline in the
deer population of the region. Hunters were driven to more
favorable hunting grounds

where they are permitted to hunt in a sportsmanlike way, untrammeled
by senseless and vicious laws. As a direct result of this, hundreds of
Adirondack guides, being deprived of their only and long accustomed
support, have been compelled by the exigencies of the case to hunt
and kill deer for the city markets, to keep their wives and children
from starvation.[57]

The journal was less charitable to the legislators following the
senate approval of a similar bill in April 1886.

Several members of the Legislatures, as Barnes, Hadley and Palmer in
the Assembly, and Kellog in the Senate, had been directed by the June

mountain mutton hotel keepers to repeal the law; they had practically no volition in the matter; they were told what to do, and when the time came they obeyed orders and did it. Mr. Husted took care to appoint the right kind of a game committee. It was a body made up of wax-noses, and their plasticity was something astonishing. It was equalled only by their avidity to be humbugged.[58]

The new law was in place for the 1886 hunting season.

The renewed anti-hounding drive of *Forest and Stream* had little effect. Whereas the prohounding coalition was showing its strength for the first time, the energy of the anti-hounding force had been sapped in its initial success. But there was some doubt that the prohibition ever had the support that *Forest and Stream* imagined. In "Let the Hounds Loose," a *New York Times* correspondent suggested that three-quarters of sportsmen opposed the anti-hounding bill that had been "pushed through" the legislature by General Curtis. A Pennsylvania sportsman offered the opinion that hounding is the only true form of sport, whereas "still hunting is an abomination." Were hounding merely an efficient method for the capture of game, as some opponents argued, Indians, for whom capture was the primary objective, would surely have followed the dogs. A spokesman for the Adirondack Club reported that visits by club members had fallen off drastically because few had the strength for the twenty-five miles of walking that the still hunt required, and he imagined that the club would disband under continued prohibition. Guides, now out of work, have complained to members who, in turn, recommend the restoration of hounding. "Personally, it makes no great difference to club members, for we can go into other states for our hunting."[59]

The repeal was further aided by the statement of Dr. Samuel B. Ward who, claiming to speak for the Eastern New York Fish and Game Protection Society, argued both that hounding was the more sportsmanlike form of the hunt and that the health of the deer herd depended upon it. Without a fear of man conditioned by the hounds, the placid deer would be wiped out by still hunters.[60] The following month, the association allegedly represented by Dr. Ward denied any support for his statement and submitted the results of a poll taken among its members that demonstrated overwhelming opposition to hounding.[61] The repeal was reportedly endorsed, too, by the New York Association for the Protection of Game, following a presentation before the membership by Assemblyman Hadley.[62] The support of the association was questioned in a letter to the *Times* by W. Holberton who claimed that the membership, following a discussion on the matter, was "decidedly opposed" to hounding and "sent a resolution to that effect to the Governor at his request."[63] Finally, the legislature received a "misleading and deceptive" pamphlet from John Denny speaking for those in favor of hounding, in which the names of Robert B. Roosevelt and Henry Bergh were connected with that sentiment. Bergh responded angrily in *Forest and Stream*. Although he was concerned with the impropriety of the practice as a sport, the major emphasis of his rejection was from the standpoint of sanitation. "It is an undeniable law of nature that the treatment of an animal at or previous to its killing is imparted for good or evil to its blood and tissues."[64] Robert Roosevelt made no response to his association with the pamphlet. In fact, with Fish Commissioner Bowman, he issued a statement that a session of mid-summer hounding was beneficial, stimulating deer to breed and toning up their systems. He subsequently "explained" this theory as a joke.[65]

The controversy over hounding frequently turned on its impact on the size and health of the deer population. This led to the trading of assertions and to the marshalling of supporting evidence on each side. The major contention of the pro-hounding forces was that the size of the harvest was reduced under a regime of hounding owing to the fear instilled in the deer by years of encounters with dogs and that the breeding stock was thereby conserved. At the same time, the popularity of hounding brought tourist money into the region, employing guides who would otherwise have to turn to market hunting. "J.T.D." argued that, at $1.50 per hide plus the small gains to be made carrying venison out of the woods on his back, the guide would have a difficult time making the $3.50 per day he could earn in the employ of visiting sportsmen. He claimed that Franklin County was deprived of $50,000 in tourist revenues during the first six months of prohibition but that the deer kill had substantially increased. Although many deer were started by dogs, only one in ten ever became the target of hunters, and even fewer were killed.[66] The counter-position was summarized in "An Economic View of Hounding," in which Adirondack guide R. W. Shutts argued that hounding was the surest way to kill deer, and, because it required no skill, increased the number of hunters in the field. Dogs could be employed in any weather. They could penetrate the deepest cover and take advantage of the deer's instinct to escape to water, where they were easy prey. Further, hounders killed a greater proportion of does than did other hunters for two reasons: first, because the females are weaker and sooner take to water and, second, because they tended to linger near water and so they were initially more vulnerable. Shutts claimed that three does were killed for every two bucks, and to this must be added a factor for the sterility induced by fright. Finally, once hounding is allowed during any portion of the year, the dogs will be used at other times.[67]

The law of 1886 easily weathered eight years of attacks, but, in 1894, the Boone and Crockett Club initiated a concerted effort to remove the hounds.[68] Bills introduced by the club in 1895 made little progress, but in 1896 the cause acquired a new strength and respectability. This was owing in part to the publicity received by the report of the State Fish and Game Commission that, of 4,900 deer killed in the 1895 season, over half had been hounded. The pro-hounding forces found themselves on the defensive, arguing not that a large and healthy population had permitted a large harvest but that the figures had been faked by resort owners and guides to attract sportsmen to the region. The detailed reporting practices of the state failed to uphold the contention. The governor now publically supported the need for change.[69]

Bills prohibiting hounding were introduced into the senate

by George R. Malby of St. Lawrence County, where hounding had been prohibited for years, and into the assembly by William Cary Sanger of Oneida. The same arguments were paraded before the legislature as had been for a decade or more. Charles H. Bennett, a Racquette Lake hotel owner, spent the legislative session in Albany and saw "a large number of the members of the Legislature and expressed to them his views and those of the Adirondack guides." The guides, who made their living from the woods and were "interested even more deeply in preserving the deer supply than the city sportsmen," continued to support hounding; wise city hunters should join them, for "not one city man in a thousand could stand the hardships of a single day's" still hunt. Ban hounding and the guides, who would see their rights as having been infringed, would become the enemies of game protection.[70] Madison Grant discounted Bennett as "simply a lobbyist for the hotel men who do not care whether there are any deer or not" and ridiculed his assessment of the city sportsman's still-hunting abilities. New York City Controller Fitch reported from the Adirondacks that guides in his part of the woods were "intelligent men" and oppose hounding.[71] In each chamber, the fish and game committees added a limited period of hounding, and, although the original text was restored on the floor of the senate, the assembly retained the amended version. Without time for a conference, Malby reluctantly accepted the assembly version as a step in the right direction.[72]

The hounding cause was finally lost in 1897. Sanger having traveled widely in the Adirondacks, claimed to have found a growing sentiment among guides and hotel keepers in favor of the absolute prohibition of hounding and jacking. The two-week period did not merit the expense of keeping hounds for the remaining fifty weeks.[73] *The New York Times* and the State Fisheries, Game and Forest Commission protectors separately recommended the ban.[74] W. H. Bowman, secretary of the Adirondack League Club, wrote, on behalf of the membership, that "much as we love fun, we believe it is public policy to stop [hounding] altogether for a term of years."[75] The legislature enacted a five-year prohibition to take effect June 1, 1897. Early first season reports showed a considerable decline in the deer kill, and, by the third season of prohibition, the Fisheries, Game, and Forest Commission could report a 50 percent increase in Adirondack deer. The prohibition "is growing in favor every year, and should remain permanently on the statute books."[76] In 1901, the prohibition was made permanent.

Restricting the Sale of Game

Perhaps the most controversial elements of the game laws in the late nineteenth century regulated the possession and sale of captured game. It seemed reasonable to game merchants, as well as in their interest, that all legally captured game should be allowed unrestricted sale. In theory, no sportsman would have objected. The logical place to regulate the hunt is at the meeting of the hunter and the hunted—by controlling time and place of capture, technology, and bag sizes. But regulation in the field is expensive, and, when game laws are unenforced, unregulated sale creates obvious problems. The development of efficient cold storage technologies in the 1870s and 1880s permitted sale throughout the year; but legal sale in the closed season encouraged illegal capture in the closed season because it is difficult to distinguish a deer killed in November from one killed in February, or a deer killed in Maine from one killed in New Hampshire, particularly once it has been butchered for sale.

Enforcement in the field represented a serious commitment of state government resources, which was slow in coming. But even with a force of salaried protectors who diligently sought out violators, evidence was difficult to secure and rural juries were at best unsympathetic to enforcement efforts. As late as 1890, a *Times* correspondent suggested that the district attornies of some Adirondack counties "would much rather have a murder trial on their hands than one game-law case."[77] While the hunt was widespread, sale occurred, for the most part, at specific and known points and within the territory of the relatively strong urban game clubs. Fine-splitting arrangements encouraged citizen vigilence; evidence was much easier to secure; and, consequently, cases left less discretion for urban juries, who were already sympathetic to the law. The restrictions on sale had the additional benefit for sportsmen of placing few limits on their own behavior. Thus, sportsmen increasingly turned their attention toward restricting the sale of game as a means to reduce the hunt.[78] Game dealers saw no reason why they should bear the burden of the state's inability to enforce its hunting regulations. They claimed to favor sound game laws and to oppose only those provisions that were "needless, tyrannical, and oppressive."[79] Knapp and Van Nostrand, a large New York City game merchant, claimed to speak for the trade in reporting that "all reputable dealers in game desire its preservation as much as conscientious sportsmen. It is against their interest to destroy any line of goods handled by them."[80]

The scarcity of game and the influence of sportsmen finally achieved total bans on the sale of game in the twentieth century, but the 1880s and 1890s, particularly in the states that supported large markets, were strewn with overturned legislation that first prohibited and then allowed the sale of game at various times under various circumstances. In January 1885, New York game dealers organized to attack what they regarded as the absurdities of the law prohibiting sale and possession of game after certain dates. "According to the law, if a man buys a dozen quail on January 31, he must cook them and eat them for dinner on the same day. If he keeps six of them for dinner on the next day, he is liable to be fined $6 and imprisoned for six days. Now this can't be right." For N. R. French, elected chairman of the group, "The food question is far more important than the question of sport."[81]

But even the ideals of sport went unrepresented in the law. Game dealer N. Durham asserted that the game laws were made "for the pleasure of dudes who spend their vacation in the Adirondacks in August and shoot does with udders full of milk and fawns running by their sides."[82] Durham's outspokenness was surely not unrelated to his arrest by the New York Association for the Protection of Game for illegally placing in storage with the Washington Street Refrigerator Warehouse Company certain commissioned quail for which he was unable to command the owner's reserve price. Charles Whitehead, long-time counsel for the association, agreed that "there are some provisions of the game laws which would be better if modified" but asserted that game dealers wanted modifications "to be sufficiently radical to allow them to sweep the game of the country into their pockets."[83]

The conflict grew in complexity in consideration of interstate game flows. Eastern markets were regularly fed by game from the west.[84] Some states whose game regularly supplied distant urban markets sought to forbid its export, but this restriction met with limited success at least in part owing to the abundance that made long-distance transportation profitable. Other states expressed no legislative concern over the loss of their own game to markets elsewhere, but the inflow and sale of this game proved troublesome to the marketing state in that it provided a cover under which local hunting might continue. State laws that prohibited the importation of game legally captured elsewhere failed the test of constitutionality but were ultimately upheld by the Lacey Act of 1900.

This mismatch between hunting and sale periods in the marketing state and hunting periods in states of great abundance proved a continual source of conflict. For New York dealers in 1891, the prohibition of the sale of venison after December 15 left no room for the sale of prime northwestern meat, and they consequently favored a total prohibition of deer hunting within the state in exchange for a sale season extending from November 1 to March 1.[85] But New York sportsmen, unwilling to give up their own late summer hunting and fearing that the deep winter deer-hunting techniques of crusting and yarding would be favored, prevented the change.

Chicago dealers had identified a different problem in the previous decade. The short marketing period at home permitted game that would have been sold in Chicago to find its way to the markets of New York and Philadelphia. They claimed that in December eastern dealers flooded the west with circulars inviting the shipment of game after the close of the Illinois market. To rectify this "injustice," Chicago dealers sought to extend the Illinois sale period through February 1 and at the same time to encourage eastern states to cut their sale periods to end on the same date. [86] The move to implement these changes was organized through the Sportsmen's and Game Dealers' Association of Chicago for the Protection of Game. The association had the initial support of Nicholas Rowe, editor of the *American Field*, as well as of a number of sportsmen. Of its two major purposes—the enforcement of the game law and the extension of the marketing period—the latter came rapidly to dominate, and the spirit of cooperation waned. Lacking the support of sportsmen, the dealers went directly to the legislature. *Forest and Stream* saw no justification for the change outside of the selfishness of marketmen. Neither did the legislature. [87]

A more serious challenge to Illinois sportsmen came in 1895 when the State Game Warden offered a bill that provided for the sale of game during the period of lawful capture but ignored entirely the issue of sale during other periods. The resistance of sportsmen was countered by the charge that they wanted only laws that served their own interests. *Forest and Stream* responded: "Of course they do. If sportsmen did not want game laws for sportsmen there would be no game laws, nor game, nor wardens to take care of the game at a salarly of $1500."[88] At the joint legislative committee hearings on the bill, Senator Berry questioned G. W. Barnett, a representative of the game dealers, and

forced [him] to admit that the law would open the market for the sale of the game of other States the year round, that it would endanger Illinois game, which could not be detected from foreign game, and that it would throw the burden of proof on the State, whereas, under the old law, the burden of proof was on the dealer.[89]

The bill was defeated. Its replacement was drafted by the Illinois Sportsmen's Association and eventually received the support of the state game warden. It prohibited the sale of Illinois game at all times and allowed the sale of out-of-state game only between November 15 and February 1.[90]

A similar controversy developed at about the same time in New York. The 1892 game laws prohibited the possession and sale of birds between January 1 and September 1, whether or not they were lawfully killed in another state. Several arrests for illegal possession and sale provided a rallying point for the American Association for the Protection of Game, Game Dealers, and Consumers which raised $5,000 to defend the New York Refrigerating Company and challenge the law. *The New York Times* saw this as a test "by the game dealers, the express companies, prominent restauranteurs, and hotel men, which, it is expected, will cause a revision by the next Legislature of the laws relating to the sale of game." L. D. Huntington of the state fish commission implied support for the change by announcing that "if there is anything unjust about the game laws of this State, it will be well to discover the injustice as soon as possible, so that the Legislature may amend the laws."[91]

The change indeed came with the 1895 game law revisions and was embodied in what became popularly known by its legal title, Section 249. The proposed version of the section stated:

No person shall be deemed to have violated any law or ordinance by reason of his selling, exposing for sale, transporting or possessing, or attempting so to do, the body, or a part of the body of any wild animal or bird in the close season for such animal or bird, provided it be proved by him that said wild animal was killed outside of this state.[92]

The bill finally enacted contained a version of this proposal which prohibited the sale of game captured within 300 miles of the borders of New York State, and it required freight bills as proof of legal importation.

The New York State Association for the Protection of Fish and Game, whose influence, following a major reorganization in 1890, was "felt in Albany as it was never felt before," was astounded by the new law. In a speech before the American Fisheries Society, Frank J. Amsden, president of the association, outlined the obvious objections to the provision. Its major effect would be

to foster and encourage crime—to put New York in the position of a fence, a receiver of stolen goods.... It says to the market-shooter,

"Go to our sister States, shoot their game in season and out of season, invoice it and ship it to the old Empire State and we will help you to dispose of your unlawfully gotten plunder." And further, it says to those of the same disposition as to our own State—and there are about as many of them—"If you can get game out of season without being caught by the protectors, box it up tight and mark it eggs or dried apples, or by some other deceptive name; we will take care of it, and when it has been mixed up with Pennsylvania or Michigan game the difference cannot be told, for the invoice of your fellow market-hunter of Pennsylvania or Michigan will cover it all.

He went on to question the legislative practices that "in some surreptitious manner and at an hour when it was impossible to correct it" enabled the disastrous section to be slipped into the act.[93]

Edward P. Doyle, secretary of the state fish commission, spoke in defense of the process and the law. An association of dealers, "composed of prominent men of this city," sent a delegation to Albany and wrote a bill introduced by Assemblyman Wilkes. The bill was reported unanimously by committee, but the chairman called in Senator Guy, "who is familiar with game laws," to modify the bill so as to protect New York game by inserting the three hundred mile clause. He concluded by asserting that the law was enforceable, and that unless and until there was evidence to the contrary, all sportsmen must assume the law will be observed.[94]

Forest and Stream, upset when the bill was introduced and greatly agitated when it was delivered for the governor's signature, was adamant when he signed the bill that it be immediately repealed. "The law is a disgrace to New York, a rank injustice to other States, a menace to the game supply, and a statute which should be rescinded: In the name of the Sportsmen of the country at large the *Forest and Stream* declares war upon the Donaldson refrigerator law; and it will not give up the campaign until the law shall have been repealed."[95]

The New York association geared up for what the *Tribune* billed as "A Bitter Game Law Fight." Robert B. Lawrence, secretary of the association, feared that "the marketmen have gained their victory and they will spend their last dollar to keep it. They look at the selling of game in the closed season as a vested right now." President Amsden outlined the strategy in a circular distributed to fish and game clubs throughout the state.

Please call a special meeting of your club; lay the matter before them; appoint an influential committee to wait on all candidates for your locality, both for the Senate and the Assembly; explain the matter to them and get their pledges to support bill for repeal [of the section] and follow it up after election by seeing the successful candidates and securing renewals of their pledges.[96]

In 1896, the New York City association, in a conciliatory mood, met with representatives of the game dealers of the city to work out a compromise for repealing Section 249. The resulting proposal failed to gain the support of the state association and made no headway in the legislature.[97] *Forest and Stream* suggested that the city association did not favor the measure either, but, convinced that Mr. Gilman, lobbyist for the market men, "owned the legislative committees," the organization accepted the proposal as the best possible outcome. The journal hoped that the association would call Gilman's bluff and change the view in Albany that "compromise, and not repeal, is the expedient action."[98]

The state association's direct attack on Section 249 initially failed because of an unfortunate link to an unimportant provision regulating saltwater fishing. In 1898, the association initiated a new effort in separate legislation, which was given a better chance. *Forest and Stream* blamed the persistence of the section on the previous inactivity of sportsmen and urged every New York citizen to write his assemblyman.[99] The passage of the repeal brought great acclaim for the efforts of Fish Commissioner Charles Babcock, whose "unfaltering determination and personal work" overcame the opposition in committee of Senator Higbie, "a stone wall in the way of right game laws," and Senator Platt, whose own law firm represented the game dealers before the legislature.[100] When the legislation was signed by Governor Black, *Forest and Stream* gave credit to President Gavitt and his associates on the law committee of the state association who had lobbied consistently.[101] But credit must be shared with George Shields, editor of *Recreation* magazine who, with several other members of the League of American Sportsmen, appeared before legislative committees and sent a brief opposing the law to the governor.[102]

The logical extension of increased restrictions on the sale of game was its absolute prohibition. As late as 1889, however, in response to a suggestion to this effect, *Forest and Stream* claimed that sale would continue as long as there was game to sell and that any attempt to further that cause through the journal was but wasted ink.[103] Yet it was just five years later that the journal proclaimed its "no sale plank" favoring the closing of all game markets.[104] While the journal claimed an enthusiastic reception to the proposal, there were some, in addition to game dealers, who opposed it. The Michigan game warden, for example, claimed in 1894 that such a provision would limit access to game to those with the time and money to hunt, whereas the resource belonged to all in common.[105] In a similar vein, Governor Hastings of Pennsylvania, in vetoing an 1895 bill prohibiting the sale of ruffed grouse, claimed that such a measure would divide the population into two groups: the group that could not eat grouse because it could not buy it and the group that could eat grouse because it could kill it.[106]

Forest and Stream saw the issue as one meriting little serious discussion. The United States was not like Britain, where game was reared in large quantities for hunting as in "a giant poultry farm." Instead, the supply of game was the result of the natural increase of wild populations and was not large enough to sustain the demands of both the sportsman and the game dealer. One of them must surrender to the other, and both qualitative and quantitative arguments would determine the victor.

That interest must give way which is of least advantage to the community, and that one must be preserved which is of paramount public importance. This is to say that the game must be saved for the enjoyment and benefit of those who pursue it for the sake of the pursuit. A grouse which gives a man a holiday afield is worth more to the community than a grouse snared or shot for the market stalls. [107]

Beyond this, "it is the protection of the rights of a hundred amateur shooters against the claim of the market killer; the privileges of the thousand against the ten, of the community against a class, and, as we have said, an extremely small and selfish class at that." [108] A number of states had accepted this judgment by the turn of the century.

The Introduction of Hunting Licenses

The hunting license, for residents and nonresidents alike, was long opposed by sportsmen because it denied free access to the common heritage in wild animals and, in the case of nonresident licenses, because it was simply unneighborly. *Forest and Stream* perhaps overstated the case in 1894:

Passports are so essential in Russia that Russian lawyers define a man as being made up of three parts—a body, a soul and a passport. If non-resident shooting and fishing legislation goes on in this country the time will come when the Russian definition of a man will be the American definition of a sportsman. [109]

Three years later, the Illinois Sportsmen's Association decreed that nonresident discriminations are "unfriendly, unsportsmanlike, un-American and wrong." [110] *Western Field and Stream*, in complaining of a nonresident license in North Dakota, asserted that "in this state [Minnesota] we have no non-resident laws, which in our view are contrary to that spirit of hospitality which imbues the true sportsman's creed." [111] Some opposed license legislation on more practical grounds. For a state such as Maine, into which nonresidents brought considerable revenues, it might make more sense to pay for the patronage of tourist hunters. [112]

The license was not universally opposed by sportsmen, however. The wealthy, as exhibited by their purchase or lease of private sporting preserves, favored exclusive access to game resources. License fees, in reducing the effective demand for hunting opportunities, might therefore find support. One Bostonian suggested that a proposed five-dollar fee for a Maine nonresident license was insufficient, and that a twenty-five or fifty-dollar license would better keep the "rabble" out of the state.[113]

While there was considerable debate on the matter of discrimination against hunters from other states, there was less debate on discrimination against hunters from other countries. In 1882, a letter to *Forest and Stream* suggested a plan for taxing foreign sportsmen in order to hire game wardens. The wealthy visitors, it was explained, would not be deterred by a small fee, but, even if they were, it would be no loss because the "universal latent spirit of toadyism in the American character," causing small towns to be proud of the presence of foreign royalty, had to be overcome.[114] By the turn of the century, the major concern was not the wealthy who came to the United States to hunt, but the poor, especially southern Europeans, who came to the United States to live. William Hornaday was among the most outspoken critics of their alleged behavior upon arrival.

On account of the now-accursed land-of-liberty idea, every foreigner who sails past the statue on Bedloe's Island and lands on our liberty-ridden shore, is firmly convinced that *now, at last*, he can do as he pleases! And as one of his first ways in which to show his independence in the Land of Easy Marks, he buys a gun and goes out to shoot "free game."

... Italians are pouring into America in a steady stream. They are strong, prolific, persistent and of tireless energy. New York City now contains 340,000 of them.... Wherever they settle, their tendency is to root out the native American and take his place and his income. Toward wildlife the Italian laborer is a human mongoose.... To our song birds he is literally a "pestilence that walketh at noonday."

. . . Let every state and province in America look out sharply for the bird-killing foreigner; for sooner or later, he will surely attack your wildlife.[115]

Recreation proposed a gun tax to deal with this "army of Italians, Bohemians, Polanders, etc."

Even a small gun license would stop a great many of these men from shooting or from carrying guns. It is well known that a dog license of $1 rids any town or city of many of its worthless curs, as soon as enacted and enforced. So a low gun license would rid every city of many of its irresponsible shooters.[116]

It was sentiment of this kind that led Pennsylvania in 1903 to pass the nation's first bill discriminating against aliens in their access to game. In 1909, unnaturalized foreign-born citizens were prohibited altogether from possessing shotguns or rifles.[117]

The extent to which both resident and nonresident licenses were adopted by 1901, in spite of the initial opposition of sportsmen, suggests a change in attitudes and the growing importance of financial considerations.[118] State level enforcement and management programs were expensive, and, as one sportsman phrased it, "It is the old and approved principle that those who dance pay the fiddler."[119] The state ownership doctrine allowed that residents and nonresidents pay different fees for the same dance. The early records on license sales indicate that few nonresident permits were issued. Michigan, in the years 1895-1900, sold over 75,000 resident, but only 304 nonresident permits. Wisconsin sold about the same number of resident, but even fewer nonresident permits per year in the years 1897-98.[120] The Wisconsin State Fish and Game Warden felt that the imposition of nonresident licenses had caused a decline in the number of nonresident visitors of three to five thousand per year. But, given that the state was surrounded by states which similarly discriminated against nonresidents, such a dramatic abandonment of Wisconsin game covers was unlikely. More probably, inadequate enforcement permitted a large number of hunters to enter the state without the purchase of a license.[121] Observers had perhaps been long overestimating the number of out of state hunters as a way of "explaining" the depletion of wildlife.

THE BROADENING OF SUPPORT

By the late nineteenth century, a history of wildlife decline had become part of the popular consciousness. Inexhaustability, never a reality, was no longer a widely accepted myth. The rapid declines of the passenger pigeon and the bison were but the most dramatic examples of man's impact on the natural world; the English sparrow had replaced the bluebird and other familiar species in the garden; some watercourses were polluted beyond their abilities to support at least the more noticeable varieties of life; and the parade of birds atop ladies' heads served as a constant reminder of the destruction that must have been left behind. The popular education offered by wide media coverage of these changes was reinforced by the growing exposure to wildlife in zoos and manageries, by a bloom of popular natural histories, and by travel and recreation that offered to thousands a firsthand view of the natural world. Science acquired a well-defined disciplinary structure supported by university departments and by professional associations that promoted organized inquiry through professional meetings and publications. A proliferation of new commissions, bureaus, and agencies marked the expanding scientific horizons of the federal government, and a growing cadre of scientific managers promoted broad developments in the theory and politics of timber, water, and grazing lands management.

This popular interest in the natural world crept slowly into a wildlife law that had historically given primary attention to game animals, furbearers, and predators. Although some protection was already afforded song and insectivorous birds, significant recognition of the wider scope of wildlife management awaited, and continues to await, the organization of special interests. These interests were not allied with sportsmen on every issue, but their general influence and political clout gave a tremendous boost to the cause of preserving populations of wild animals.

The humane movement, indefatigably led by Henry Bergh, president and founder of the American Society for the Prevention of Cruelty to Animals, offered a particularly pointed reminder that increased interest in animals was not always increased support for sportsmen. Of particular distaste to Bergh was the pigeon shoot, popular pastime of gunners, and primary attraction at any number of sportsmen's meetings.[122] Bergh intervened in New York as early as 1872. The sporting press branded him an eccentric, but feared that his brand of eccentricity might generate political support. If he carried his point in New York, the *American Sportsman* argued, his success "will be eagerly seized upon by old fogy humanitarians everywhere, and used to further their crazy schemes for preventing the popular and excellent sport of trap shooting."[123] There was no place for considerations of cruelty. "It being admitted that man has right of sovereignty over the lower animals, Mr. Bergh and the humanitarians desire to restrain the exercise of that right."[124] *Forest and Stream* assumed a more utilitarian stance:

In pigeon practice, the death of each bird ought to bring some compensating benefit to the contestants, whether in reward of merit, the pleasure of honorable emulation, or in improved accuracy. We never could bring ourselves to believe that pigeons were created for the express purpose of being shot from the trap, although they seem in this way to serve men best. . . . So long as it is more important that our citizens should become expert in the use of arms than that the lives of thousands of pigeons should be saved, so long shall we defend the practice of trap-shooting.[125]

Continued pressure throughout the decade brought concessions from New York sportsmen in the operation of the shoot and outright prohibitions in at least three other states.[126] *Forest and Stream* feared that laws prohibiting all shooting of game might follow. "Then the S.P.C.A. will turn its attention to the suffering Walnut Hill rifle targets."[127] More important in limiting the hunt than moral outrage was the scarcity of live targets. Existing mechanical substitutes, usually glass balls released from spring traps, were rapidly improved and widely adopted. *Forest and Stream* endorsed the glass ball as a cheap and effective substitute for live birds, and one immune to Bergh's attacks.[128]

Among the organizations more closely linked with the sportsmen's cause were the American Ornithological Union (AOU) and the Audubon Society. The AOU, modeled on the British Ornithological Union, grew out of the Nuttall Ornithological Club in 1883. Among its members were the best-known naturalists and ornithologists of the day: D. G. Elliot, A. K. Fisher, Elliot Coues, Robert Ridgeway, C. Hart Merriam, William Brewster, George B. Sennett, J. A. Allen, George Bird Grinnell, C. E. Bendire, and Spencer F. Baird. Six investigative committees, to consider classification and nomenclature, migration, anatomy, oology, the English sparrow, and faunal areas, were supplemented in 1884 by a seventh to consider the general protection of bird populations.[129] The first report of this new committee appeared as a special supplement to *Science* magazine on February 26, 1886.[130] "With a desire to bring about more intelligent, uniform and desirable legislation for the protection everywhere, and at all times, of all birds not properly to be regarded as game-birds, ... " the committee proposed a "model law" for consideration by state legislatures. The proposal offered full protection to non-game birds, their eggs and nests. Collection for scientific purposes was to be allowed only after careful review by a responsible state organization or official.[131] The English sparrow was alone among nongame birds excepted from protection. An important feature of this proposal was its independence from legislation affecting game birds. The reason for this

is mainly to avoid conflict of interests respecting such legislation, which is more or less sure to follow in any attempt at combined legislation respecting all birds in one act. Sportsmen's clubs and game-protective associations in attempting to provide proper game-laws often find strong opponents in the game-dealers and market-gunners, who often succeed in defeating judicious legislation. If all birds are treated under the same act, attempts to improve the portions of such acts as relate to useful birds are often prevented through opposition to certain clauses of the game-sections obnoxious to pot-hunters and game-dealers.[132]

The "model law" was immediately adopted by New York, followed by Pennsylvania in 1889 and by several other states before the turn of the century. Even where the law itself was not adopted, nongame birds were afforded greatly increased protection.

The market demand for these birds arose largely from the millinery trade.[133] While a change in the laws making bird plumage more difficult to acquire was important in reducing the annual harvest, this required a level of enforcement that was not forthcoming in these early years of concern. Some felt that more success would result from consumer education. The AOU Committee on Bird Protection observed that "of all the means that may be devised for checking the present wholesale bird slaughter, the awakening of a proper public sentiment cannot fail of being the most powerful."[134]

It was in this spirit that *Forest and Stream* announced the formation of the Audubon Society in February 1886. The journal offered a free membership card to anyone taking one or more of three pledges: to prevent, to the extent possible, "(1) the killing of any wild birds not used for food; (2) the destruction of nests or eggs of any wild bird, and (3) the wearing of feathers as ornaments for dress."[135] By June, 10,000 "bird lovers," many of them school children, had enrolled. Membership rose to 16,000 by December and to a reported peak of 37,400 by August 1887. In 1889, under financial pressures, *Forest and Stream* abandoned the program.[136]

The popular sentiment favoring the protection of nongame birds, uncovered by the Audubon Society, continued to spread. The revival of feathered fashions in the 1890s provided the catalyst for the formation of a new breed of Audubon clubs, first in Massachusetts and Pennsylvania in 1896, and in more than a dozen other states by 1900. In 1901, the existing clubs joined to create the National Committee of Audubon Societies, which, with limited success, coordinated member clubs, and, in 1905, the thirty-five state societies formally incorporated as the National Association of Audubon Societies for the Protection of Wild Birds and Animals.[137]

Concern for the well-being of birds whose nesting grounds were pillaged to meet the whims of fashion had itself become fashionable. This fashion was promoted by a series of "inspirational volumes in the guise of gentle bird lore for children and the fair sex," represented by the works of Olive Thorne Miller, Florence Merriam, Mable Osgood Wright, and Neltje Blanchan.[138] T. S. Palmer, acting chief of the Division of Biological Survey of the U.S. Department of Agriculture, attributed the "extraordinary popular interest in bird study which has developed in the past few years," to the introduction of nature study in the common schools, and the efforts of the Audubon Societies in the cause of bird protection." He reports that 70,000 texts on birds had been sold by New York and Boston publishers during the previous six years and that his department had distributed 200,000 publications during the same period.[139]

To the extent that the protection of birds followed the restraint of the millinery trade, Massachusetts was a leader in protection. In 1897, U.S. Senator George F. Hoar was able to secure the passage of a bill prohibiting the possession or wearing of the feathers of any birds protected by the laws of the state. He drew up a petition from the song birds of the state, complete with their "signatures," which was read to the legislature by A. S. Roe, state senator from Worcester. "I thought it might, perhaps, strike the Legislature of Massachusetts and the public more impressively than a sober argument."[140] The petition read:

We the song birds of Massachusetts and their playfellows, make this our humble petition:

We know more about you than you think we do. We know how good you are. We have hopped about the roofs and looked in at the windows of the houses you have built for the poor and sick and hungry people and little lame and deaf and blind children. We have built our nests in the trees and sung many a song as we flew about the gardens and parks you have made so beautiful for your own children, especially your poor children, to play in.

. . . We are Americans just as you are. Some of us, like some of you, came from across the great sea, but most of the birds like us have lived here a long while; and birds like us welcomed your fathers when they came here many years ago. Our fathers and mothers have always done their best to please your fathers and mothers.

Now we have a sad story to tell you. Thoughtless or bad people are trying to destroy us. They kill us because our feathers are beautiful. Even pretty and sweet girls, who we should think would be our best friends, kill our brothers and children so that they may wear their plumage on their hats. Sometimes people kill us from mere wantonness. . . .

Now we humbly pray that you will stop all this, and will save us from this sad fate. You have already made a law that no one shall kill a harmless song-bird or destroy our nests or our eggs. Will you please to make another one that no one shall wear our feathers, so that no one will kill us to get them? We want them all ourselves. . . . [141]

The New York Times was skeptical of anyone's ability to legislate fashion and said of Hoar's petition that the "sentiment of this work of literary art is impeccable, while its style and quality would do credit to any pupil of a Massachusetts girls' school."[142] That the measure easily passed is indicative of the force of sentiment, properly expressed, in the face of more conventional interests.

ENFORCEMENT OF LAW

The wisest laws will utterly fail if they are neither enforced nor widely supported. Throughout most of the nineteenth century, enforcement was the weakest link in the development of protective legislation for wildlife. It is costly, and conviction often required a guilty verdict by peers who may have had nothing but contempt for the law itself. Through 1885, which may be taken as the initiation of concerted efforts by states to enforce their own game laws, the most effective enforcement agencies were the sportsmen's clubs whose members brought violations to the attention of officials authorized to prosecute them.

The most effective clubs were those located near urban game markets where potential violations were confined and enforcement costs were small. The New York Association for the Protection of Game kept a particularly vigilant watch on the New York City markets as early as 1870. In December of that year, at the monthly meeting of the association, President Royal Phelps remarked that the good work of the club had had a great effect in halting the sale of illegal game and in encouraging high-class eating establishments to obey the law. "[Now] to invite guests to eat game out of season is to insult them."[143] Using its membership fees and dues, the club employed a staff of detectives to roam the market stalls in search of violations. In 1873, for example, the association brought twenty-seven suits for violations of the game laws, all but three of which were decided in favor of the club.[144] On March 1, 1877, the first day of the closed season, three detectives filed twenty-one complaints from which eighteen suits resulted. According to Charles Whitehead, counsel to the association, this represented the "usual number of delinquents engaged in selling game contrary to law."[145] The most notable early victories of the association were in the cases against dealers Middleton and Carmen for possession of one hundred speckled trout in violation of the law, settled in the amount of $2,500 (or $25 per fish as specified by law), and against Joseph Racey for possession of quail beyond the legal date of sale. This latter case was appealed to the state appeals court, which upheld the conviction in a landmark decision supporting state authority to limit the sale of game.[146]

Many other local game clubs worked to enforce the law, though only occasionally with the vigor of the New York association. Clubs in rural areas, concerned more with the violation of hunting than of marketing provisions, found enforcement in the field difficult, if for no other reason than that the suspect was armed. A nonaggressive violator could easily claim to be in search of a species in season at the time. The enforcement of a prohibition of Sunday shooting was less problematical. A Staten Island club organized in 1874 for this purpose and, with the assistance of the county police, arrested thirty-four hunters the first Sunday out. By the third Sunday, only three hunters could be found and, thereafter, the club could report that "the crack of a gun mingling with the sound of the different church bells is a thing of the past."[147] But such successes were rare. More often, widespread opposition to the game laws and a fear of reprisals that might have followed assistance in enforcement frustrated the work of the local game clubs. In 1877, the Hartford Game Club distributed leaflets throughout eastern Connecticut summarizing the game laws and offering ten-dollar rewards for information on violations. Some letters were received, but "in each case the witnesses stated that they were so fearful of having their barns burned or their cattle or horses maimed or killed, if known to have given this information, that they were unwilling to aid in convicting the offender."[148]

The system of voluntary enforcement by local clubs potentially offered the vigilance of thousands of sportsmen anxious to halt the destructive practices of the market hunter, but it offered also the temptation to relax that vigilance to accommodate their own hunt. One scenario viewed the local sportsman's club as a means to establish a monopoly over access to game within a region by prosecuting nonmembers for violations while respecting the rights of members to engage in hunting at their own discretion.[149] The game laws of the 1870s had been written largely by sportsmen, with little organized opposition, and were enforced entirely by sportsmen. This combination might invite arbitrary limitation of access to a privileged class from the cities. Legitimacy required the commitment of public funds to the maintenance of a force of game protectors.[150]

In 1874, the *American Sportsman* recommended that each state appoint a game commissioner to secure adequate legislation and coordinate the efforts of sportsmen's clubs in its enforcement.[151] The following year, the journal supported the appointment of a United States game commissioner to head a new organization modeled after the U.S. Fish Commission, which had been created in 1871.[152] *Forest and Stream*, in its turn, suggested that each state appoint detectives, under the supervision of district deputies and a chief game commissioner, to roam areas most frequented by hunters.[153] Similar proposals regularly emerged from sportsmen's conventions and interstate conferences.

Beginning with New Hampshire and California in 1878, states gradually accepted the responsibility for managing game resources.[154] Most agencies were created initially at the state level, but some had their origins in local efforts which can be traced back to colonial reeves and wardens. Thus, the modern Connecticut system began with an 1883 provision that the selectmen of each town were to appoint two or more wardens to assist in detecting and prosecuting game law violations. In 1889, control was elevated to the county level by the appointment of eight wardens, one for each county. It was not until 1895, however, that a state-level office was created. Likewise, in Illinois, protective legislation began with the 1885 appointment of three wardens, one each for the three largest cities, who were charged with the enforcement of regulations concerning the sale of game. It was not until 1899 that a state game commissioner was appointed, who, in turn, was authorized to appoint one warden for each congressional district within the state.

The creation of the state game commissions by no means offered immediate relief for wildlife. The annual *Reports* of the New Hampshire commission, the first to turn its attention formally to game, do not mention wildlife until 1884, six years after the shift. The report of that year referred to numerous complaints that "rough grouse" were being snared contrary to law but observed that the commission had "not succeeded in obtaining evidence sufficient to convict the guilty parties as yet." The first convictions for snaring grouse, killing deer, fishing in closed season, and possessing short lobsters were reported in 1886. Limited funding was at least one cause of minimal efforts. By 1890, the annual appropriation to support agents in the field was $600, of which only $314.10 was spent. In spite of this, the commission could report that "the lives of hundreds of deer were saved. After it became known that detectives were appointed, the crust hunters in almost all places stopped their work of killing deer." More than twenty poachers were arrested and fined.[155]

Limited funds and half-hearted efforts were second to the tremendous opposition to regulation, which often made early enforcement futile. State control from afar was viewed as an unwarranted interference with customary liberties and demonstrated a failure to recognize the needs and preferences of local populations. The North Dakota fish and game commissioner, in his first report, observed that "the game laws of North Dakota cannot be executed, public sentiment will not justify it."[156] The Vermont commission, in its 1895-96 report, asserted that "it has been the policy of the commission to enforce all laws in the chapter of game laws, regardless of their popularity," as if that had never before been the case.[157] The report from Michigan for the same years recognized the impossibility of uniform and strict enforcement. Deer violations among residents "who really need the meat" were settled with a verbal reprimand, whereas violations by "people in good circumstances" were prosecuted to the full extent of the law.[158] The New Jersey wardens were plagued with bad publicity about their extreme vigilance. Part of this, the commission confessed, resulted from the unauthorized enforcement by individuals who brought violations to trial to collect that portion of the penalties prescribed by law. Certain justices, it appears, were able to secure a cut for their cooperation in these matters. At the same time, the commission was careful to point out that its own agents engaged in no malicious enforcement or harassment. In particular, in disclaiming a story about the arrest of four girls for possessing caged robins, the report proudly states, "Our records show that no girl or woman was ever arrested for violation of the fish and game laws."[159]

A closer examination of the structure of protection in several states demonstrates a wide range of styles. In New York, the

state-level warden system was initiated by an 1880 act author-
izing the governor to appoint eight fish and game protectors at
an annual salary of $300. The appointments were made with-
out regard to geography, and certain areas of the state, notably
New York City and Long Island, were unprotected.[160] In 1883,
the number of protectors was increased to sixteen, to be
appointed by district.[161] The protectors served under the state
fish commissioner until 1888 when the office of chief protector
was created. The fish commission, charged with the appointment of the chief protector, seemed reluctant to act. *Forest and
Stream* located the source of this reluctance in the control that
Governor Hill maintained over the tenure of the commissioners. There was some uncertainty whether their terms continued
for life or whether they had just expired. The commissioners
feared that if they selected a chief protector who did not please
the governor on political grounds, he would subscribe to the
latter interpretation.[162]

Apparently in response to this indictment by the journal, the
commission, under a resolution that the office of chief protector be a nonpolitical one, selected Fred P. Drew, the eleventh
district protector.[163] Drew's two-year tenure ended with the
charge that he had been too vigilant in protecting the state's
wildlife resources. In particular, it was suggested that he kept
illegal fish out of the Fulton Market in New York City and that
this offended the fish commission, whose president, Eugene
Blackford, had a place of business there. General U. Sherman
resigned his post on the commission in protest over Drew's
removal. Blackford, in response to these charges, claimed that
Drew had been initially appointed on the recommendation of
Sherman, that he had been disliked from the start, that he
showed disrespect for the commission, that he failed to make
regular reports of his work as required by law, and that he
spent the majority of his time lobbying in Albany rather than
patrolling his district in the field. Drew's replacement, J.
Warren Pond, was conceded by *Forest and Stream* to have been
the best of the remaining protectors, and the credibility of the
commission was reestablished.[164]

In 1892, Governor Hill removed Blackford from the commission because he had supported the governor's opponent in the recent election. While he believed Blackford to be a good man for the job, "he did not propose to keep Mugwumps in office."[165] *Forest and Stream* regretted this as but one more step in the direction of converting the game protection system into a part of the political machine. "We are to have not game protectors but ward heelers; the increase of the food fish supply is to be subordinated to the satisfying of partisan greed."[166]

The political control over the game protective network carried down into lower levels. John Burnham sought the position of protector in Essex County. He first consulted with Walter Witherbee, the local political boss, who sent him for the endorsement of the county committee. The local committeeman indicated that he would be hired if he showed promise of enforcing the law in a "judicious" manner, whereupon he replied: "Doc, if you have in mind that I will be somewhere else when Walter Witherbee and his friends go hounding, you tell them I'll enforce the law damn *injudiciously*." He received instead the position of county excise commissioner.[167]

The political climate within which the commission continued to operate did not restrain individual game protectors from enforcing the law. Each was required to submit an annual summary of his activities, and the summaries were included in the biennial reports of the commission. Judging from these reports, the protectors differed greatly in the enthusiasm and dedication with which they approached their jobs. Some protectors made a monthly or even daily accounting of time and effort. Some spent well beyond the amounts that could be reimbursed. [168] Some noted extreme difficulty in obtaining convictions, even when the cases seemed ironclad. Seymour Armstrong, of the fifth district, apprehended a man whom two witnesses had just seen hounding a deer, following it into the water, and hauling it into his boat.

And yet, when Pasco, in a hang-dog manner, told the jury that he was in there on the first of July with his dog and gun *hunting* for *ginseng,* the jury by their verdict assumed to believe him. I hope that the fact that some of the members of the jury were bushwacking about with Pasco's attorney nearly all night before the verdict was rendered, had nothing to do with the decision.[169]

On a more favorable note, the report of 1888 records cases in Hamilton County in which the offenders were found guilty by their own grand juries. "Such an occurrence has hitherto been thought impossible."[170]

In New York, then, the state agency for the protection of game assumed a rather passive stance, mediating diverse interests and enforcing, to varying degrees, whatever legislation emerged. In Maine, on the other hand, where game was regarded as an important economic asset in drawing sportsmen to the state, and where political interests concerning wildlife were in considerable harmony, the commissioners of fisheries and game became strong advocates for the wise use of the state's wildlife resources. In their first report, in 1880, the commissioners stated,

It must always be borne in mind that the Commissioners of Fisheries and Game have no more to do with the sporting side of fish and game, than farmers. Our duties are the production of food. To show that we can plant an acre of water and produce as much food as from an acre of well tilled and cultured wheat. The harvesting of the crop may be sport or work, according to the means, or necessities, or taste of the reaper.[171]

The following year, the commissioners confirmed that "it has ceased to be a matter of question or experiment any longer, that a stock of both fish and game can be kept up to the full extent of the feeding power of the waters and forests of a given territory, by a stringent enforcement of the laws of protection, during their respective breeding and recuperating seasons."[172]

The game of Maine was clearly an asset that merited protection and sound management. In one early study on the value of Maine game to the state, Lucius L. Hubbard compiled ex-

penditure statistics for thirteen groups, including 185 sportsmen and 101 guides, which made seventy-five tours to Maine
between 1868 and 1881. Their total expenditure in Maine was
$13,403, of which $5,228.50 was paid to guides at the rate of
$2.75 per day, and their total kill was seven moose, four cari-
bou, and four deer. To compare the gains to the state from
sport hunting with those from market hunting, Hubbard hypothesized a party of two market hunters taking twenty moose
each, yielding 12,000 pounds of meat worth 6¢ per pound and
the hides, worth $10 each. Estimating total expenses at $60, he
calculates a net profit of $250 per hunter. Based on the sample,
tourists would bring some $37,000 into the state in killing
twenty moose. Further, whereas market hunters are not selective in their hunt, sportsmen shoot more bulls than cows.
Suppose that the twenty were not killed by market hunters
and that they were left alone for five years. Assuming that a
majority of the initial twenty were females, the population
after five years would have increased to one-hundred. A 10
percent natural decrease plus a 20 percent legal hunt leaves
seventy. If fifty were shot by sporting tourists during the five
years, twenty would remain to begin the cycle again. Meanwhile, the fifty moose would have brought a total of $90,000, or
$18,000 per year into the state. Thus, two hunters destroyed in
one month, and for $500, what would yield $18,000 per year to
the state forever.[173] The message is clear in spite of questionable calculations. A later report, by E. C. Farrington, secretary
of the Maine Sportsmen's Fish and Game Association, claimed
that the state attracted some 200,000 tourists annually, of
whom 50,000 were sportsmen spending an average of $60, for
a total of $3,000,000. Capitalized at 6 percent, the stock of game
resources within the state had a value of $50,000,000. The
assessed valuation of 9,000,000 acres of wild lands within the
state was by comparison only $17,000,000.[174]

Other states generally concerned with the value of game
resources were Vermont, New Hampshire, Michigan, Wisconsin, and Minnesota. The fish commissioner of Minnesota estimated, in 1875, that the annual value of game in that state was

$500,000. "We have something worthy of the attention of the political economist."[175] The report of the Michigan commissioner in 1876 likewise included an economic plea for the protection of wildlife:

> Our native fishes,—the Trout, the Greyling, the Bass, the Whitefish, —our wild poultry,—the deer and other wild game of our forests, what have they cost us? And how immensely have they contributed to the aggregate wealth of the people, and to the general attractiveness of the State? . . . [Many nonresidents view Michigan as the] fur, fin, and feather arcadia for the exercise of the rod and the gun. . . . Is it not, then, the province of a wise, forecasting people, to see to it that these benefits and these attractions are not diminished, but rather, if possible, increased, and their permanent safety, instead of being jeopardized, hedged about, if need be, by inexorable penal law, for their loss, if but partial even, like the loss of our forestry, would be irreparable, betokening quite unmistakably the dawning of an era of pusillanimity and decay.[176]

The calculations of the state of Maine found their way into the reports of a number of other state commissions. The Wisconsin Fish and Game Commission report for 1897-98 noted on "good authority" that the hunters and fishermen visiting Maine spent $3,000,000 each year.[177] This figure also appeared in the Vermont report of 1893-94 as did a report that New Hampshire took in some $6,000,000 annually from tourism. The commissioner saw no reason why Vermont, with the same potential, could not generate the same income by improving its protective efforts.[178] The New Hampshire report for 1896 claimed that the average deer carried out of the state by a hunter yielded that state some $150.[179] *Forest and Stream*, appealing to economic calculations such as these, encouraged western big-game states such as Idaho, Montana, and Wyoming to take a business perspective on their game resources.

> The citizens of these states should look at this matter purely from a business point of view, from the standpoint of dollars and cents. No

people in the world are keener business men than the citizens of these communities or quicker to see a business point. . . . The businessman can influence the newspapers, and in a short time public opinion can be so altered that men who have been accustomed to violate the law would no longer do so, finding themselves in danger of prosecution on the one hand and on the other without a market for their skins.[180]

The role of state fish and game agencies was new and uncertain during the nineteenth century. It would be greatly strengthened in the twentieth century as funding was guaranteed, as expertise replaced politics, and as game management techniques were devised. But the controversies would not be resolved. Over some issues, and in some states, the politics of wildlife management in the 1970s is practically indistinguishable from that nearly a century ago.[181]

THE EXPANDING FEDERAL ROLE

The outstanding features that distinguish twentieth-century wildlife management from that of the nineteenth century are the expansion in coverage from game and other commercially valuable species to broader classes of wildlife and the shift from state and local to national and international perspective and control.[182] The former change has its origins in the nineteenth-century humane and Audubon movements and reflects the evolution of preferences concerning wildlife as expressed through existing political channels. The latter change represents a shift from concerns rooted in state authority to those rooted in national authority and, more importantly, an evolution in the interpretation of the boundary that divides state from national jurisdiction. It was not the migratory patterns of waterfowl that underwent dramatic shift with the turn of the twentieth century, but rather a shift in perceptions and institutions that permitted an expanded role for the federal government in signing international treaties to manage migratory flocks, which deprived the several states of control that they had historically exercised.

The federal government had always played an important, but indirect, role in wildlife management—as the public domain shifted from public to private ownership, as Indian policies changed the face of the frontier, as jurisdiction over navigable waterways and coastal water expanded, as the U.S. Department of Agriculture sponsored investigations into wildlife on the farm and judged its ultimate status as friend of foe, and as courts assessed the constitutionality of state regulation. But prior to 1900, the federal government's direct role in regulating wildlife was extremely limited. The major legislation enacted during the nineteenth century concerned hunting on Indian territory (1834),[183] the sale and possession of game in Washington, D.C. (1878 and 1899),[184] and the protection of wildlife in Yellowstone (1894) and Mt. Ranier (1899) National Parks.[185]

The range of increased federal activity was limited. The common understanding, as well as the Supreme Court's ruling that ownership of game was held by the several states, suggested that a national game law would be unlikely to emerge. U.S. Representative George G. Vest, an important conservationist and the leader of the fight to save Yellowstone Park during its early years, considered a national law "so abhorrent to my ideas of the structure and powers of our Government that it requires some exertion on my part to consider the matter seriously."[186] A national law had virtually no support even for migratory species, which, owing to their interstate travels, might appear to better justify federal intervention. The National Association for the Protection of Game, Birds and Fish observed in 1886 that although the effective control of waterfowl would be greatly improved by such activity, "it is quite clear that Congress has no power to legislate on the subject."[187]

Several problems in state regulation suggested the propriety of federal intervention could justification be found in the enumerated powers of the Constitution. The first was the states' difficulty in protecting local wildlife when game from other states was imported for local sale. But animals legally captured in other jurisdictions and legally exported had entered the stream of interstate commerce and could not be denied entry by individual states. Second, the notorious rise from obscurity of the English sparrow and the starling and the fear of repeat performances by other exotic species suggested the careful monitoring of imported plant and animal material, a task quite beyond the states. Finally, the popular work of the U.S. Fish Commission in the propagation and distribution of food and game fishes suggested a parallel role for a national game commission in supporting state management efforts.

Congress, in a drawn-out, four-year process, replete with the usual political machinations, considered and finally accepted these responsibilities with the passage of the Lacey Act. The act was the descendant of bills introduced in the House by John Lacey of Iowa on July 1, 1897, and in the Senate by George Hoar of Massachusetts on March 14, 1898.[188] The Lacey bill proposed to broaden the duties of the U.S. Fish Commission to include "the propagation, distribution, transportation, introduction, and restoration of game birds and other wild birds useful to man," recognizing that the laws of the individual states would always govern the purchase or capture of birds for such purposes. The Committee on Merchant Marine and Fisheries recommended passage without amendment, arguing that "birds introduced by the National Government ... would be protected by quite a different state of public opinion from that which has prevailed as to birds introduced by private individuals or shooting clubs."[189] But the House, sympathetic to the view that "there is something more important in these war times than hatching ducks and goslings," refused to take up the measure.[190] Six months later, the bill was reconsidered, passed, and referred to the Senate.[191]

Senator Hoar's bill sought the protection of song birds through the prohibition of interstate transportation of "birds, feathers, or parts of birds to be used or sold." Hoar had secured the passage of a similar bill in his home state, largely on the emotion generated by a petition "signed" by the song birds of Massachusetts, which he submitted to the legislature.

A request was made that the same document be read before the Senate, after which the bill was quickly passed and sent to the House.[192]

Upon receipt of Lacey's bill by the Senate, Hoar, fearing that his own measures would not pass the House, recommended amending the legislation by substituting his proposal for its entire text, thus guaranteeing a conference. It was then suggested that the two bills instead be joined, to which Hoar had no objections. The combined bill passed and was sent to the House, on January 12, 1899. Lacey felt that the Senate version was too broad in its restraint of trade, but this matter was little debated. Most of the concern was focused on the importance of halting the further decimation of bird populations and on the propriety of the original Lacey proposal.[193] Broadening the fish commission's role seemed to open the door to the establishment of a government aviary in every congressional district. Furthermore, the line between state and federal authority was not similarly drawn for fish as for birds because navigable waterways were the clear property of the federal government but most bird habitat was not. Consideration was postponed, pending the recommendations of a conference committee.[194] The consensus of the committee was that the jurisdiction of the fish commission ought not to be enlarged. This was in no small part in recognition of that agency's opposition to its proposed role. In response, the Senate passed an amended version of the bill from which Lacey's provisions had been stricken, and the matter was returned to conference, from which it would not emerge.[195]

Lacey reintroduced his legislation, now with the Department of Agriculture's Division of Biological Survey saddled with the responsibilities of propagation and distribution. The department, which had not yet publically agreed to its proposed role, suggested that "preservation" replace "propagation" and that "the means to be adopted to secure the preservation of birds be left to the discretion of the department." More importantly, the department drafted a new section of text which would place the introduction of exotic birds and animals under U.S. Department of Agriculture control.[196] Lacey accepted the modifications in revised legislation. The House Committee on Interstate and Foreign Commerce further amended the proposal by adding a provision which spoke to the concerns of the Teller and the Collum and White bills.[197] Based on the language of the Wilson Original Package Act of 1890, which regulated the interstate flow of alcoholic beverages, the section subjected wildlife imported from another state to the laws of the local state, original packaging as proof of origin notwithstanding.[198]

Support for the legislation grew quickly among sportsmen and in the sporting press. Notably active was George Shields, who personally wrote to all the members of Congress and could report to Lacey by March 12 that "at least 150 members have promised unequivocal and positive support . . . [and] only 2 objected to the measure in any way."[199] In debate on the floor of the House, the bill was further amended by subjecting foreign game to state law and, to break finally the opposition of milliners, by excluding barnyard fowl from the provisions of the act. There was little major opposition, and the bill was passed on April 30. The Senate passed the bill, without interference by Hoar and without debate, on May 18, and it was signed into law by President McKinley on May 25, 1900.[200]

The passage of the Lacey Act neither assured the future of the nation's wild animals nor streamlined the regulatory process. There was now another level of control that provided another foundation for the construction of vested interests. The landscape continued to change, largely outside of the scope of any existing regulatory authority. Legislation based on incomplete, if not inaccurate, information about wildlife populations continued to be passed, often to be repealed and then passed again. The species that persisted were largely those able to coexist with land distrubances and those requiring habitats not in demand for other uses. But despite the immediate prospects for most wild animals, the organizational forms and institutional controls that emerged in the final third of the nineteenth century have thoroughly structured, and will continue to structure, the complex social process of allocating competing claims to common property in wildlife.

NOTES

1. A printed copy of this letter, dated December 18, 1870, presumably prepared for wider distribution, may be found with the papers of the New York Association for the Protection of Game in the archives of the New York State Library at Albany. On the later hostilities between the city and state organizations, see note 22 below.

2. *American Sportsman* 7 (January 15, 1876), 248.

3. *Forest and Stream* 16 (May 26, 1881), 323.

4. *Shooting and Fishing* 14 (June 1, 1893), 105-6.

5. *Science* 7 (February 26, 1886), 203. The article went on to note that all other birds "should be left to the care of bird-lovers and humanitarians. . . ." The pages in this supplement (191-205) constituted the first report of the Committee on Bird Protection of the American Ornithological Union.

6. *Forest and Stream* 2 (March 12, 1874), 74.

7. *Forest and Stream* 44 (February 16, 1895), 121.

8. William T. Hornaday, *Our Vanishing Wildlife* (New York: The New York Zoological Society, 1913), pp. ix, 204.

9. This entrenchment is supported by hunting- and fishing-license revenues and by taxes on sporting equipment that are earmarked for game management and habitat improvement. Only recently have taxes on the accoutrements of nonconsumptive uses of wildlife (for example, binoculars and birdseed) been considered as a source of funds for the management of nongame species. For a popular account of hunter domination, see Cleveland Amory, *Man Kind? Our Incredible War on Wildlife* (New York: Harper and Row, 1974), pp. 19-194.

10. See Samuel P. Hays, *Conservation and the Gospel of Efficiency: The Progressive Conservation Movement, 1890-1920* (Cambridge, Mass: Harvard University Press, 1959).

11. The first important work on the subject was Aldo Leopold's *Game Management* (New York: Charles Scribner's Sons, 1933).

12. Connecticut, *Report of the State Commissioner of Fisheries and Game* (1895/96), p. 9; New York, *Report of the Commissioners of Fisheries, Game and Forests for 1895*, p. 10, Michigan, *Report of the State Board of Fish Commissioners* (1893/94), p. 9.

13. *Forest and Stream* 40 (February 16, 1893), 133. The Michigan fish commission suggested that the fish and game laws of any state should be "of such compact sense and form that they may be pasted in the crown of your hat." [Michigan, *Report of the State Board of Fish Commissioners* (1876), 41].

14. *Forest and Stream* 40 (March 16, 1893), 225.

15. *Forest and Stream* 28 (April 28, 1887), 300-301; 30 (May 3, 1888), 291; 32 (May 23, 1889), 355; 48 (April 3, 1897), 268.

16. These numbers are derived from the "Schedule of Laws Repealed" that is appended to Ch. 19, "Forest, Fish, and Game Law" [*Consolidated Laws of the State of New York* 5 vols. (Albany: J. B. Lyon Company, State Printers, 1909), II, 1155-72.] Intrastate conflicts which appeared to derive from regional differences in the quality of game populations or in climate, rather than in the preferences of hunters, were occasionally resolved by granting some authority over game to localities or by separately considering the game of several regions. New Jersey, for example, was divided, in 1888, into a northern and a southern region, each with its own seasons for important species. The regions were merged in 1889 and re-created in 1895.

17. See Chapter 5 for a discussion of the uniformity of the law.

18. *Forest and Stream* 12 (February 27, 1879), 71-72. Under the plan, the seasons for deer, ruffed grouse, pinnated grouse, and woodcock would be closed between January 1 and September 1; for quail between January 1 and October 1, and for turkey between February 1 and September 1. The sale of all of these species would be prohibited after January 15.

19. See Map 4. See also *Forest and Stream* 48 (March 27, 1897), 241. The original scheme included Arizona in the western region.

20. *Western Field and Stream* 2 (July 1897), 80.

21. *The New York Times*, September 27, 1896, Sunday Magazine, p. 10. By the Migratory Bird Treaty concluded with Great Britain in 1916, the federal government assumed the authority to regulate access to certain species of birds by all residents of all states and territories. It is of interest to compare the extent of uniformity achieved under this broad mandate with that sought by sportsmen during an earlier period in the absence of centralized authority. The general terms of the treaty established a closed season on migratory game birds between March 10 and September 1 (except that in the Atlantic States and provinces north of the Chesapeake Bay, the closed season for shore birds was established between February 1 and August 15) and provided that open seasons must be limited to three and a half months of the remaining period, to be determined severally by the contracting powers. Other migratory nongame birds were to be protected year round except for certain hunting allowed to Eskimos and Indians. The treaty further provided that a number of species be fully protected for periods of five to ten years.

The specific regulations proclaimed by the president under the treaty, to take effect July 31, 1918, divided the territory of the United States into five regions, each with its own open season for waterfowl and shore birds. Opening dates ranged from September 1 to November 1 and closing dates ranged from November 30 to January 31. In addition, daily bag limits were established for all regions: ducks, 25 aggregate; geese, 8 aggregate; brant, 8 aggregate; plover and yellow-legs, 15 aggregate; Wilson snipe, 25; woodcock, 6; and doves, 25 aggregate.

By September 1935, there had been thirty-four amendments to the regulations accompanying the treaty. Their influence in terms of unification was mixed. By that date, waterfowl regulations had been reduced to three sets, two for the contiguous states and one for Alaska. The seasons were both shorter and more uniform than in the original regulations. Woodcock, on the other hand, was given seven different open seasons, beginning between October 1 and December 1 and ending between October 31 and December 31. Bag limits were generally reduced. Ducks were limited to 10 in the aggregate, geese and brant to 4 aggregate, rails to 15, woodcock to 4, mourning doves to 20, and Wilson's snipe to 15 [U.S. Department of Agriculture, Bureau of Biological Survey, Service and Regulatory Announcements No. 23 (August 26, 1918) and No. 83 (September 1935)].

22. *The New York Times*, November 10, 1874, p. 2; *Forest and Stream* 3 (November 12, 1874), 216. The exclusion of the New York City association's delegation from the state convention held in Oswego in July 1874 on uncertain technical grounds prevented that club's proposal for a national convention from being presented for consideration. This is reflective of the general tensions that existed between the two organizations and that aid in the explanation of Hallock's contempt [*Forest and Stream* 2 (July 2, 1874), 328].

The state association's call, it seems, was initially motivated by the *American Sportsman*, *Forest and Stream's* early competitor for the attention of eastern sportsmen [*American Sportsman* 3 (March 7, 1874), 360]. It was not until the New York City association had fully accepted the national convention called by the New York State association that *American Sportsman* was willing to recognize the importance of the city association. [*American Sportsman* 5 (November 14, 1874), 104]. Whatever antagonism remained in 1876 between the state and city organizations was greatly reduced by the election of Col. Alfred Wagstaff of the city association to the vice-presidency of the state association. [*Forest and Stream* 6 (June 1, 1876), 271]. See also Theodore W. Cart, "The Struggle for Wildlife Protection in the United States, 1870-1900:

Attitudes and Events Leading to the Lacey Act" (Ph.D. diss., University of North Carolina, 1971), pp. 92-96.

23. *American Sportsman* 4 (September 19, 1874), 392.

24. *American Sportsman* 4 (September 12, 1874), 373, 380. On the property status of dogs, see U.S. Department of Agriculture, *Report of the Commissioner of Agriculture for the Year 1863*, "Dogs and Dog Laws," by J. R. Dodge (Washington, D.C.: G.P.O., 1863), pp. 450-63; and George Putnam Smith, *The Law of Field Sports* (New York: Orange Judd Publishing Co., 1886), pp. 19-32.

25. *American Sportsman* 6 (June 19, 1875), 179.

26. *Forest and Stream* 7 (September 7, 1876), 73; 10 (June 20, 1878), 387.

27. In 1875, there was an attempt to form an international association for wildlife protection. It was organized to gather information and was not to compete with the national organization already in existence. The attempt was noble in purpose and was supported by a host of prominent sportsmen and scientists including Robert B. Roosevelt, president of the American Fish Culturists' Association, Elliot Coues of the Smithsonian Institution, and the fish commissioners of New York, Massachusetts, Maryland, Michigan and Canada. Standing committees were formed on nomenclature and geographical distribution, on habits, and on the law, each of which was to report findings as they were compiled. No findings were reported in *Forest and Stream* [*Forest and Stream* 4 (April 8, 1875), 136; 4 (April 29, 1875), 186-87; 4 (May 27, 1875), 250; 5 (November 18, 1875), 249-50; 5 (December 9, 1875), 282; and 5 (January 20, 1876), 377-78].

28. *Forest and Stream* 22 (February 31, 1884), 66.

29. *Forest and Stream* 25 (August 13, 1885), 41.

30. *Forest and Stream* 25 (October 8, 1885), 206-7.

31. *Forest and Stream* 33 (January 2, 1890), 465.

32. *Forest and Stream* 40 (June 12, 1893), 511.

33. *Forest and Stream* 41 (September 30, 1893), 269.

34. See Cart, "The Struggle for Wildlife Protection," pp. 100-102. The association's greatest notoriety came with the support and publicity it gave to an alleged mass destruction of Alaskan duck eggs for the confectionary and photographic trades. The story, which turned out to be largely, if not wholly, without foundation was not put to rest before Congress was encouraged to halt the nefarious trade by preventing the importation of game bird eggs in 1894 (Tariff Act of 1894, 28 Stat. 540). The ban was renewed in 1897 (30 Stat. 197). These prohibitions proved a burden to sportsmen's organizations and state

commissions interested in obtaining eggs for restocking game covers with imported species. On the incident in general, see "The Great Duck Egg Fake," *Forest and Stream* 44 (June 22, 1895), 503-5; George Bird Grinnell, *American Duck Shooting* (New York: Forest and Stream Publishing Company, 1901), pp. 576-82. Jenks Cameron, *The Bureau of Biological Survey: Its History, Activities and Organization* (Baltimore: Johns Hopkins University Press, 1929), pp. 82-83.

35. Grinnell, experienced western traveler, authority on the Plains Indians and friend and confidant of Roosevelt, joined *Forest and Stream* in 1876 when he took over the editorship of the Natural History column from Ernest Ingersoll, with whom Hallock had become increasingly dissatisfied. During the next several years, Grinnell and his father purchased significant stock in the company. By 1879, "Mr. Hallock, president of the company, had become more and more eccentric, drinking heavily and neglecting his duties to the paper. . . . " E. R. Wilbur, treasurer and major stockholder, suggested to Grinnell that he move to New York (from New Haven) and become the president of the Forest and Stream Publishing Company. Grinnell agreed, and Hallock, under the pressure of the major stockholders, reluctantly gave up his position and sold his stock [John F. Reiger, ed. *Passing of the Great West: Selected Papers of George Bird Grinnell* (New York: Winchester Press, 1972), pp. 143-44]. The journal which prospered under Grinnell's management, deteriorated rapidly after his departure in 1911, became a monthly in 1915, and merged with *Field and Stream* in 1930.

Grinnell and Roosevelt first became acquainted over the former's critical review of Roosevelt's *Hunting Trips of a Ranchman*, which appeared in *Forest and Stream* in July 1885. They became fast friends [John Franklin Reiger, "George Bird Grinnell and the Development of American Conservation" Ph.D. diss., Northwestern University, 1970), pp. 163-70. A major theme of this study is the important role that Grinnell played in preparing Roosevelt for the leadership that he was to provide for the progressive conservation movement after the turn of the century]. It was Grinnell who appeared as the major driving force behind the Boone and Crockett Club. Explorer and naturalist Charles Sheldon wrote that the club "has been *George Bird Grinnell* from its founding. All its books, its works, its soundness, have been due to his unflagging work and interest and knowledge" (quoted from a letter to Redmond Cross, in Reiger, "George Bird Grinnell," p. 179).

36. See James B. Trefethen, *Crusade for Wildlife—Highlights in Conservation History* (Harrisburg and New York: The Stackpole Com-

pany and the Boone and Crockett Club, 1961), for a detailed account of the formation and activities of the club. See also George Bird Grinnell, "Brief History of the Boone and Crockett Club," in George Bird Grinnell, ed., *Hunting at High Altitudes* (New York: Harper & Brothers, 1913).

37. Articles II, III, and V of the Constitution of the Boone and Crockett Club (Trefethen, *Crusade for Wildlife*, pp. 356-57). With the increased mobility of the membership and with the increased scarcity of American big game, the constitution was amended in the twentieth century to allow the substitution of two males of foreign big game for two of the three kills required for membership.

38. Theodore Roosevelt, "The Boone and Crockett Club," *Harper's Weekly* 37 (March 18, 1893), 267.

39. A discussion of the controversy over hounding may be found in Ch. 6, text at notes 48-76. The protection of the game of Yellowstone came in 1894 (28 Stat. 73). See Trefethen, *American Crusade*, pp. 76-90, and Reiger, *American Sportsmen*, pp. 114-41. In addition, Grinnell attributed to the efforts of the club the passage of the Forest Act of 1891, the creation of the New York Zoological Society, the passage of the Alaska Game Law of 1902, the establishment of Glacier National Park, and the general movement of the federal government toward the establishment of game preserves ("Brief History of the Boone and Crockett Club," passim).

40. The idea for the organization was suggested by a reader in the issue of October 1897. George Oliver Shields (1846-1925) had written extensively on sporting topics under his pen-name, Coquina, previous to the establishment of *Recreation: A Monthly Magazine Devoted to Everything that the Name Implies*, in 1894. He lost the journal in 1905 to personal bankruptcy, whereupon he published *Shields' Magazine* under the sponsorship of the New York Zoological Society until 1912. From 1897 to 1902 he headed the Campfire Club of America, which had been established by William Hornaday.

41. *Recreation* 7 (October 1897), 266-67. See *Recreation* 7 (December 1897), 465-68, for reader response.

42. *Recreation* 9 (December 1898), 465.

43. *Recreation* 12 (February 1900), 139-40.

44. See text at notes 91-102 on the New York law and text at notes 188-200 on the Lacey Act, and see Cart, "The Struggle for Wildlife Protection," pp. 175-81.

45. U.S. Department of Agriculture, Biological Survey Bulletin No. 16, pp. 41-42.

46. *Forest and Stream* 28 (May 19, 1887), 371.

47. *Forest and Stream* 28 (May 12, 1887), 341. The existence of this provision was blamed for the early failures of the Massachusetts Fish and Game Protective Association's efforts to stock regions of the state with prairie chickens, some 750 of which had been liberated by 1893 [ibid., 36 (March 26, 1891), 188; 42 (January 22, 1894), 71].

48. Massachusetts, *Report of the Commissioners on Inland Fisheries and Game* (1887), p. 32.

49. Charles Hallock, *The Sportsman's Gazeteer and General Guide*, 5th ed. (New York: Forest and Stream Publishing Company, 1879), p. 82.

50. The practice had been prohibited statewide at times in the past and in certain counties during this period. The game law of 1871, for example, prohibited hounding in any county north of the Mohawk River, but the provision was repealed in 1879.

51. The total prohibition against hounding was weakened by an amendment offered by Senator James Otis which excluded Suffolk County. Deer there, Otis told his colleagues, ought to be shaken up by the hounds. Presently, they were so tame that one came up to a farmer, smelled him, and "was killed by the farmer with an axe." The Senate was immediately persuaded. Not so Henry Bergh, the indefatigable champion of animal welfare, who wrote to Otis that the action of the Senate was in "flagrant opposition to the instincts of advanced civilization" *The New York Times*, April 25, 1885, p. 3., and May 5, 1885, p. 7.

52. *Forest and Stream* 24 (April 30, 1885), 265.

53. *The New York Times*, January 8, 1886, p. 3.

54. *Forest and Stream* 25 (December 17, 1885), 405.

55. *Forest and Stream* 25 (January 4, 1886), 501.

56. *Forest and Stream* 26 (February 25, 1886), 81.

57. *Forest and Stream* 26 (March 4, 1886), 105.

58. *Forest and Stream* 26 (April 15, 1886), 221.

59. *The New York Times*, February 8, 1886, p. 5, and February 10, 1886, p. 5.

60. As support, Ward claimed that the deer herd had grown under the regime of limited hounding in effect from 1879 to 1885. *Forest and Stream* 26 (February 11, 1886), 49.

61. *Forest and Stream* (March 25, 1886), 161.

62. *Forest and Stream* (February 11, 1886), 41; *The New York Times*, February 9, 1886, p. 2.

63. *Forest and Stream* 26 (February 18, 1886), 69; *The New York Times*, February 15, 1886, p. 2.

64. *Forest and Stream* 26 (February 18, 1886), 69. Bergh, founder of

the ASPCA in 1866 and the best known spokesman for the rights of animals until his death in 1888, often directed his attention toward sportsmen and market hunters. The most heated debate concerned the propriety of pigeon-shooting contests which were the highlight of sportsmen's conventions in the 1870s. See text at notes 122-28.

65. *Forest and Stream* 29 (December 8, 1887), and (December 29, 1887), 441.

66. *The New York Times*, January 8, 1886, p. 3. Another letter to the *Times* suggested that one deer in four started by dogs falls before the hunter's gun (August 18, 1889, p. 13). Dr. Ward suggested that the figure was one in three [*Forest and Stream* 26 (February 18, 1886), 70].

67. *Forest and Stream* (February 25, 1886), 88-89.

68. Early in 1894, Roosevelt wrote Grinnell: "Don't you think the executive committee plus Madison Grant, might try this year to put a complete stop to hounding in the Adirondacks? Appear before the Legislature, I mean. I wish to see the [Boone and Crockett] Club do something" (cited in Trefethen, *An American Crusade*, p. 111).

69. *The New York Times*, February 10, 1896, p. 9; February 12, 1896, p. 7; *Report of the Commissioners of Fisheries, Game and Forest for 1895*, pp. 159-253.

70. *The New York Times*, February 10, 1896, p. 9.

71. *The New York Times*, February 1, 1896, p. 9.

72. Madison Grant, "Reform in New York Game Laws," *Harper's Weekly* 40 (October 3, 1896), 978.

73. William Cary Sanger, "The Adirondack Deer Law," in George Bird Grinnell and Charles Sheldon, eds., *Hunting and Conservation* (New Haven: Yale University Press, 1925), p. 276.

74. *The New York Times*, January 9, 1897, p. 4; February 21, 1897, p. 3.

75. *The New York Times*, March 16, 1897, p. 12.

76. *The New York Times*, October 19, 1897, p. 7; *Report of the Commissioners of Fisheries, Game and Forests for 1899*, p. 84.

77. *The New York Times*, March 16, 1890, p. 17.

78. As late as 1915, the Boone and Crockett Club considered this legislative form as but a second best solution. Non-sale laws were "a confession that the laws for killing game are not enforced. They are justified on the ground of expediency, as the only effective method of preventing traffic in game" (Report of the Game Preservation Committee, p. 53).

79. *Forest and Stream* 23 (January 22, 1885), 506.

80. *Forest and Stream* 35 (January 1, 1891), 476.

81. New York *Tribune*, January 9, 1885, p. 1.

82. New York *Tribune*, January 9, 1885, p. 1. This was not an isolated charge. The New York laws, in particular, seemed geared to the summer vacation plans of certain sportsmen who wished to hunt and fish for trout at the same time. "The peculiar game law of this State was framed at the solicitation of certain influential parties whose season for pleasure and recreation comes in the summer."(*The New York Times*, December 23, 1884, p. 6).

83. New York *Tribune*, January 10, 1885, p. 8.

84. Some markets, particularly Boston and Washington, D.C., because of their extended sale periods, became known as "dumping grounds" for game illegally captured in or illegally exported from other states. As an indication of the extent of interstate trade, the Massachusetts Fish and Game Protective Association claimed in 1896 that 90-95 percent of the game sold in the Boston market originated outside of Massachusetts. [*Forest and Stream* 46 (March 14, 1896), 214]. The same claim was made for the New York City market in 1885. [*Forest and Stream* 23 (January 22, 1885), 506].

85. *Forest and Stream* 35 (January 1, 1891), 476.

86. *Forest and Stream* 18 (February 2, 1882), 8; *American Field* 18 (January 14, 1882), 48.

87. *Forest and Stream* 20 (February 22, 1883), 61. Organizations similar to this one emerged in other marketing cities. In New York, the American Association for the Protection of Game, Game Dealers and Consumers was formed in 1885 [*Forest and Stream* 23 (January 15, 1885), 481].

88. *Forest and Stream* 44 (February 9, 1895), 101.

89. *Forest and Stream* 44 (March 2, 1895), 169.

90. *Forest and Stream* 44 (March 9, 1895), 181.

91. *The New York Times*, June 1, 1892, p. 9.

92. *Forest and Stream* 44 (May 11, 1895), 363.

93. *Forest and Stream* 45 (July 20, 1895), 53. The association's lobbyist in Albany relaxed his vigilance having been assured that the section had been killed, only to find it a part of the final legislation that had been rushed through both houses (*The New York Times*, June 13, 1895, p. 16).

94. *Forest and Stream* 45 (July 20, 1895, 53. Enforcement was indeed a problem. Chief Inspector Pond reported in the first hunting season following passage that violations were numerous due precisely to the 300-mile-limit law "secured by the market men" which provided a cover for illegal hunting in New York (*The New York Times*, February 10, 1896, p. 9).

95. *Forest and Stream* 44 (February 23, 1895), 141; (May 25, 1895, 407; (June 15, 1895), 481. Charles N. Cuthbert, treasurer and counsel for the state association, and Thomas Whitehead, counsel for the New York City association, wrote Governor Morton urging that he withhold his signature from the Donaldson bill, but the session had concluded and Morton signed the bill to capture what he considered its beneficial portions (New York *Tribune*, June 2, 1895, p. 22, and December 8, 1895, p. 17).

96. New York *Tribune*, December 8, 1895, p. 17.

97. *Forest and Stream* 47 (November 21, 1896), 410.

98. *Forest and Stream* 47 (March 20, 1897), 221.

99. *Forest and Stream* 50 (March 12, 1898), 201. J. Warren Pond, chief protector for the State of New York, offered repeated support for repeal. (*Report of the Commissioners of Fisheries, Game and Forests for 1896*, p. 253; *Report* for 1897, p. 146).

100. *Forest and Stream* 50 (April 2, 1898), 261.

101. *Forest and Stream* 50 (April 16, 1898), 301.

102. *Recreation* 8 (May 1898), 401.

103. *Forest and Stream* 32 (February 14, 1889), 61.

104. *Forest and Stream* 42 (February 3, 1894), 89.

105. *Forest and Stream* 44 (February 2, 1895), 90-91; Michigan, *Report of the State Game and Fish Warden* (1893/94), p. 10.

106. *Forest and Stream* 44 (June 1, 1895), 437. The opposition was not limited to those who saw distributional implications. The debate over sale again raised the question of the state ownership doctrine that legitimized such controls over hunter behavior. The battle standard was carried by Dwight W. Huntington, editor of the *Amateur Sportsman* (later the *Game Breeder*) from 1898 to 1912. In his view, the disappearance of wild animals in the United States resulted from the state ownership doctrine and the consequent absence of management incentives offered to individuals. In 1909, Huntington launched a search for the "discoverer" of the doctrine [*Amateur Sportsman* 41 (August 1909), 3]. Huntington proposed a model "Breeders' Law," which granted ownership of animals and fish to landowners or lessees who artificially propagated, reared, and/or protected them from natural enemies.

107. *Forest and Stream* 42 (February 10, 1894), 111.

108. *Forest and Stream* 44 (June 1, 1895), 437.

109. *Forest and Stream* 43 (July 14, 1894), 23.

110. *Forest and Stream* 47 (February 27, 1897), 167.

111. *Western Field and Stream* 2 (August 1897), 99.

112. Letter to *Forest and Stream* 40 (February 16, 1893), 133. Another objection centered on the dubious constitutionality of the matter. Though the courts did not deal squarely with the issue in the nineteenth century, *Allen* v. *Wyckoff,* 48 N.J. Law Rep. 90 (1886) and *In Re Eberle*, 98 Fed. 295 (1899), are generally taken as upholding the constitutionality of nonresident licenses. See Chapter 5, note 62, for a discussion of these cases.

113. *Forest and Stream* 40 (March 9, 1893), 208.

114. *Forest and Stream* 18 (February 2, 1882), 10; ibid. (February 9, 1882), 27.

115. William Hornaday, *Our Vanishing Wildlife*, pp. 101-2.

116. *Recreation* 8 (May 1898), 401. Italian hunters remain under attack in Europe owing to their predeliction for species that migrate between northern Europe and Africa. It was estimated that 1.6 million Italian hunters killed 150 million nongame birds in 1971. (*The New York Times*, April 16, 1972, section 1, p. 3).

117. The 1903 statute required a ten-dollar county license for nonresidents and unnaturalized residents.

118. See Map 3 and U.S. Department of Agriculture, Bureau of Biological Survey, *Hunting Licenses; Their History, Objects, and Limitations*, by T. S. Palmer, Bulletin No. 19 (Washington, D.C.: G.P.O., 1904).

119. *Forest and Stream* 13 (January 22, 1880), 1010.

120. U.S. Department of Agriculture, Biological Survey Bulletin 19, p. 40; Wisconsin, *Report of the Commissioners of Fisheries and State Fish and Game Warden* (1897/98), pp. 88-89.

121. This is supported by statistics that show that Wisconsin sold over twenty times the number of nonresident and over five times the number of resident licenses in 1903 as in 1898. Surely the increase in the number of hunters was not of this magnitude. [U.S. Department of Agriculture, Biological Survey, Circular 54, "Statistics of Hunting Licenses" (Washington, D.C.: G.P.O., 1906)]. Johnson reports that some North Dakota hunters were not aware that licenses were required until 1910, more than a decade after their introduction [Morris D. Johnson, *Feathers from the Prairie: A Short History of Upland Game Birds* (Bismarck: North Dakota Game and Fish Department, 1964), pp. 22-23].

122. Birds were released from traps, and shooters would attempt to down the birds before they could escape. Generally, each contestant shot at 10 birds, the tied shooters in each round moving on to further contests until the entire field was ranked. The larger shoots might

consume 25,000 birds, so that several hundred thousand birds would easily have been required to supply the needs of various pigeon shoots that took place during an average year in the 1870s.

123. *American Sportsman* 1 (February 1872), 4.

124. *American Sportsman* 8 (August 19, 1876), 330.

125. *Forest and Stream* 3 (August 22, 1974), 41.

126. The New York sportsmen agreed, for example, that shooters would be stationed near the traps to prevent the escape of wounded birds, that killed birds would be used as human food, and that only wild birds would be shot [*American Sportsman* 4 (May 9, 1874), 93].

127. *Forest and Stream* 12 (April 17, 1879), 216.

128. *Forest and Stream* 12 (July 19, 1879), 470. For Henry Bergh's endorsement, see *Forest and Stream* 7 (August 24, 1876), 40. For Adam Bogardus's account of his development of the "rough surface" glass ball, see Bogardus, *Field, Cover, and Trap Shooting*, 2d ed. (New York: Published by the author, 1878), pp. 380-89.

129. The ambition of the Union was in part responsible for the federal government's role in economic ornithology. See Chapter 3, note 62.

130. *Science* 7 (February 26, 1886), 191-205. The supplement included articles by J. A. Allen on "The Present Wholesale Destruction of Bird-Life in the United States," by William Dutcher on "Destruction of Bird-Life in the Vicinity of New York," by George B. Sennett on "Destruction of the Eggs of Birds for Food," as well as anonymous pieces entitled "The Destruction of Birds for Millinery Purposes," "The Relation of Birds to Agriculture," "Bird Laws," and "An Appeal to the Women of the Country in Behalf of the Birds."

131. Such a procedure was by no means a guarantee of protection. A similar law in Massachusetts was of little value. "If a local party has found the law standing in the way of his pleasure, all that has been necessary was to write to President Seelye of Amherst College, and a permit to shoot birds has been immediately forthcoming" [*Forest and Stream*, 20 (June 14, 1883), 387].

132. *Science* 7 (February 26, 1886), 203-4.

133. See Chapter 3, text at notes 43-46.

134. *Science* 7 (February 26, 1886), 195. In 1886, the New Hampshire Board of Fish and Game Commissioners issued an invitation to "the Ladies of the State" to aid in "abolishing the cruel practice of sacrificing so many beautiful and useful birds for securing ornaments to gratify a depraved and wicked taste." [New Hampshire, *Report of the Fish and Game Commissioners* (1886), p. 11].

135. *Forest and Stream* 26 (February 11, 1886), 41; 26 (March 18, 1886), 141.

136. *Forest and Stream* 26 (June 24, 1886), 425; 27 (December 2, 1886), 361; 28 (April 7, 1887), 221; and 29 (August 18, 1887), 61; Peter Matthiessen, *Wildlife in America* (New York: The Viking Press, 1959), p. 167.

137. The association took its current name, National Audubon Society, in 1940. Several state societies remain autonomous. On the history of the society, see Carl W. Buchheister and Frank Graham, Jr., "From the Swamps and Back—A Concise and Candid History of the Audubon Movement," *Audubon* 75 (January 1973), 4-45.

138. Mattheissen, *Wildlife in America*, p. 178; see Cart, "The Struggle for Wildlife Protection," pp. 114-45, passim. The cause of song and insectivorous birds was further boosted by Professor Charles A. Babcock, superintendent of schools in Oil City, Pennsylvania, who proposed, in 1894, the annual celebration of Bird Day in the public schools of the nation. See U.S. Department of Agriculture, Division of Biological Survey, Circular 17, "Bird Day in Schools," by T. S. Palmer (Washington, D.C.: G.P.O., 1896); and Charles A. Babcock, *Bird Day— How to Prepare for It* (New York: Silver, Burdett and Company, 1901).

139. *U.S.D.A. Annual Report of the Secretary, 1899* (Washington, D.C.: G.P.O., 1899), p. 64. The department saw for itself the role of coordinating the efforts of wildlife organizations by preparing a list of state commissions, sportsmens's associations, Audubon societies, League of American Sportsmen chapters, etc., for inclusion in the 1899 *Yearbook of Agriculture* (*Annual Report of the Secretary*, 1900, pp. 38-39).

140. George F. Hoar, *Autobiography of Seventy Years*, 2 vols. (New York: Charles Scribner's Sons, 1903), II, 274-77.

141. A facsimile of the petition may be found in *New England Magazine*, New Series, 16 (July 1897), 614-15. The document was also introduced into the *Congressional Record* (55th Cong. 2nd Sess., pp. 3166-67) in the debate leading to the passage of the Lacey Act.

142. *The New York Times*, July 18, 1897, p. 14.

143. *The New York Times*, December 16, 1870, p. 2.

144. *Forest and Stream* 1 (November 13, 1873), 218.

145. *Forest and Stream* 8 (April 12, 1877), 148.

146. See Chapter 5, text at notes 26-27 for a discussion of this case.

147. *American Sportsman* 5 (October 3, 1874), 10.

148. *American Sportsman* 9 (October 4, 1877), 170-71.

149. Occasionally, sportsmen felt duty-bound to report instances in which they violated, or failed to enforce, the game laws. "Medi-

cus," in describing a restaurant meal, confessed to his "inability" to enforce the law prohibiting the sale of woodcock during certain months.

> Then followed reed birds, upland plover, brown backs, and -ah, me! must I confess it, "Owls from the North." Don't suppose for a moment, gentle reader, that I am going to tell you what an "Owl from the North" is. Think of the most toothsome bird *out* of season in August, and mentally make a note of it. When they appeared, my duty as an officer of a protection society was clear, but—one of those aforesaid troublesome buts—*absolvo me*, I had seen them in the larder, and I deliberately stifled the warning voice of conscience, and ate: nay, I fairly revelled in the delicious juices. [*American Sportsman* 9 (February 3, 1877), 280].

150. In the states of New Jersey, Delaware, and North Carolina, an intermediate solution developed in which private organizations of sportsmen were given semi-public status with respect to their control over game resources. This provided some level of enforcement without public funding. See Tober, "Allocation of Wildlife Resources," pp. 392-96.

151. *American Sportsman* 4 (May 2, 1874), 72.

152. *American Sportsman* 6 (September 18, 1875), 376.

153. *Forest and Stream* 5 (December 23, 1875), 312-13.

154. See Map 4 above.

155. *Report of the Fish and Game Commissioners of New Hampshire* (Concord), 1884, p. 10; 1886, p. 10; and 1891, pp. 57-58.

156. Morris D. Johnson, *Feathers from the Priarie: A Short History of Upland Game Birds* (Bismarck: North Dakota Fish and Game Department, 1964), p. 98. In some cases, it was the game wardens who were executed. Two Maine wardens were murdered by deer hunters in 1886 [*Forest and Stream* 27 (November 11, 1886), 301]. In 1905, a warden employed by the Audubon Society to enforce the Florida game law was murdered by plume hunters. In Pennsylvania, in the single year of 1906, fourteen game protectors were shot at, of whom seven were hit, and four killed. The trial for one of these murders, which involved members of the Italian Black Hand Society, was instrumental in providing support for legislation in that state prohibiting aliens from possessing firearms (Schulz, *Conservation Law and Administration*, pp. 48-49).

157. Vermont, *Report of the Commissioners of Fisheries and Game, 1895/6*, p. 50.

158. Michigan, *Report of the State Fish and Game Warden, 1895/6* p. 9.

159. New Jersey, *Report of the State Fish and Game Commissioners, 1897* pp. 15-16.

160. *Forest and Stream* 15 (September 30, 1880), 163.

161. In spite of the enabling legislation, only thirteen districts were initially created. In 1886, the number was increased to seventeen, reduced to fifteen in 1889, raised to twenty in 1892, to thirty-eight in 1895, and to fifty in 1905.

162. *Forest and Stream* 31 (August 30, 1888), 101.

163. *Forest and Stream* 31 (September 6, 1888), 121.

164. *Forest and Stream* 35 (November 20, 1890), 351. According to the report of the commission, the change was made "for merit alone," and "for reasons satisfactory to the commission" (*Report of the Commissioners of Fisheries for 1891*, p. 20).

165. *Forest and Stream* 38 (January 14, 1892), 25.

166. *Forest and Stream* 38 (January 7, 1892), 1.

167. Marvin W. Kranz, "Pioneering in Conservation: A History of the Conservation Movement in New York State, 1865-1903" (Ph.D. diss., Syracuse University, 1961), pp. 225-26.

168. See, for example, New York, *Report of the Commissioners of Fisheries* (1886), pp. 159-60.

169. Ibid., p. 103.

170. Ibid., (1888), p. 21. The enforcement activities of some other states were recorded in admirable detail. Connecticut, for example, gained convictions for the following game law violation in the period 1899-1900: Sunday hunting, 21; hunting song birds, 4; snaring game birds, 3; dynamiting for fish, 2; taking trout less than 6 inches, 2; taking pickerel less than 6 inches, 4; killing English pheasant, 1; illegal fishing, 7; illegal hunting, 2 [*Report of the State Commissioners of Fisheries and Game* (1896/98), p. 45; ibid. (1898/1900), p. 51].

The degree of detail shown in this report and in those of New York is hardly matched by most. Ohio, for example, in the reports of its county wardens for 1886, offers the following statements in their entirety:

The law is not violated in reference to fish and game.
There has not been a general observance of the law.
No violation of the fish and game laws in my jurisdiction.
Cannot say that the law has been observed.
The law has been well observed.
The law has not been observed at all.
There has been a general observance of the law. I have removed

over 700 illegal fish nets [Ohio, *Report of the Ohio State Fish Commissioner* (1886), pp. 32-40].

171. Maine, *Report of the Commissioners of Inland Fisheries and Game* (1880), p. 5.

172. Ibid. (1881), p. 4.

173. *Forest and Stream* 18 (May 11, 1882), 284.

174. *Forest and Stream* 44 (February 9, 1895), 112.

175. Minnesota, *Report of the State Fish Commissioners* (1875), p. 22.

176. Michigan, *Report of the State Board of Fish Commissioners* (1876), p. 4.

177. Wisconsin, *Report of the Commissioners of Fisheries and State Fish and Game Warden* (1897/98), p. 92.

178. Vermont, *Report of the Commissioners of Fish and Game* (1893/94), pp. 6-7. Maine's total tourist income was estimated at $10,000,000 of which $3,000,000 was attributed to sportsmen.

179. New Hampshire, *Report of the Fish and Game Commissioners* (1896), pp. 8-9.

180. *Forest and Stream* 41 (August 5, 1893), 93.

181. In 1979, the Vermont legislature, for the first time in its history, and after a battle of many years, relinquished control over the state's deer herd to the Fish and Game Department. The key point of contention had been the propriety of an anterless season which the legislature had forbidden (except for several limited hunts in the 1960s) and which has now been implemented to the extreme displeasure of many residents and legislators. "If there's one single subject on which every single Vermonter feels he or she is an expert, it is management of the deer herd" [Brattleboro *Reformer* (November 30, 1978), p. 4]. Ernest "Stub" Earle, chairman of the House Fish and Game Committee and opponent of surrendering control, predicted a large landposting drive as a protest against the doe season. It "is the only whip the country people have." The problem "is essentially a problem of 'city versus country.' . . . City folks now dominate the Legislature, but country folks own the land on which the deer is hunted" [*Reformer* (March 15, 1979), p. 1].

182. The Endangered Species Act of 1973 defines "fish or wildlife" to include "without limitation any mammal, fish, bird . . . amphibian, reptile, mollusk, crustacean, anthropod or other invertebrate, and includes any part, product, egg, or offspring thereof, or the dead body or parts thereof" [Sec. 3 (8)]. The act also offers protection to plants. See also Council on Environmental Quality, *Evolution of National Wildlife Law*, pp. 444-51.

183. 4 Stat. 730. "That if any person, other than an Indian, shall, within the limits of any tribe with whom the United States shall have existing treaties, hunt, or trap, or take and destroy, any peltries or game, except for subsistence in the Indian country, such person shall forfeit the sum of five hundred dollars, and forfeit all the traps, guns, and ammunition in his possession, used or procured to be used for that purpose, and peltries so taken."

184. 20 Stat. 134; 30 Stat. 1012.

185. The Yellowstone Act (28 Stat. 73) provided "that all hunting, or the killing, wounding, or capturing at any time of any bird or wild animal," except to protect humans from injury or death, was prohibited. The Mt. Ranier Act (30 Stat. 993) provided protection only against "wanton destruction" and use for profit.

During the 1870s, Congress pondered the fate of the buffalo. A number of bills were introduced, and one, which provided that no one other than an Indian could kill female buffalo within the territories, nor could anyone kill more male buffalo than could be used for food or market, passed both houses but was never signed into law by President Grant. See Chapter 3, note 103.

186. *Forest and Stream* 34 (June 5, 1890), 389.

187. *Forest and Stream* 27 (August 12, 1886), 41. In congressional debate preceding the passage of the Lacey Act, Lacey explained that the legislation would not directly prohibit the taking of animals because "to do that it would become necessary to enact a national game law, which . . . would be unconstitutional" [*Congressional Record*, 56th Cong. 1st Sess. (April 30, 1900), p. 4873].

188. *Congressional Record*, 55th Cong., 1st Sess., p. 2195; 2nd sess., p. 2757. Another strand of legislation was introduced in January 1896 by Senator Henry Teller of Colorado who sought to regulate the illegal export of big game from Colorado, Utah, and Wyoming [*Congressional Record*, 54th Cong., 1st Sess. (January 23, 1896), p. 896]. The bill was not reported from committee and was reintroduced in 1897, 1898, and 1899 [*Congressional Record*, 55th Cong., 1st Sess., (July 2, 1897), p. 2189; 55th Cong., 2nd Sess. (January 18, 1898), p. 718; 56th Cong., 1st Sess. (December 15, 1899), p. 440]. Its major features were finally included in the Lacey Act. Shortly after Teller first introduced his bill, the proposal drafted by F. S. Baird of the National Game, Bird and Fish Protective Association was introduced by Congressmen Shelby Moore Cullom and George Elon White of Illinois [*Congressional Record*, 54th Cong., 1st Sess. (February 5, 1896), p. 1317; (February 7, 1896), p. 1519]. This proposal would similarly have restricted the illegal export of enumerated game species. Neither the House nor the

Senate version was reported from committee. A more detailed legislative history of the Lacey Act may be found in Cart, "The Struggle for Wildlife Protection," and in Cart, "The Lacey Act: America's First Nationwide Wildlife Statute," *Forest History* 17 (October 1973), 4-13.

189. *Congressional Record*, 55th Cong., 2nd Sess., H.R. Report 522.

190. Comment of Mr. McMillan. *Congressional Record*, 55th Cong. 2nd. Sess. (April 12, 1898), pp. 3743-44.

191. *Congressional Record.*, 55th Cong., 3rd. Sess. (December 19, 1898), pp. 317-18).

192. *Congressional Record*, 55th Cong., 2nd Sess. (March 24, 1898), pp. 3166-67. The text of the Hoar bill and other legislation may be found in Cart, "The Struggle for Wildlife Protection," Appendix I.

193. House concern for the decline in bird numbers resulted in part from the findings of William Hornaday's survey on wildlife abundance conducted for the New York Zoological Society. He mailed questionnaires to "persons competent to answer them" asking, among other things, for judgements on the abundance of birds relative to fifteen years previous. From these reports, "fully 90%" of which "bear unmistakable evidence of having been prepared with conscientious thought and care," Hornaday concluded that thirty states had witnessed a decrease in bird life averaging 46 percent and ranging from 77 percent in Florida to 10 percent in Nebraska. North Carolina, Oregon, and California had shown no change in bird populations, and Kansas, Wyoming, Washington, and Utah had shown increases.

The reliability of these figures is open to serious question. The Connecticut figure, for example, appears to have been based on two reports, for which the printed digests were as follows: "Fairfield, Mrs. Mabel Osgood Wright, 'We have not a quarter as many of these kinds (game birds, meadow-nesting song birds, and marsh birds) as ten years ago.' Portland, John H. Sage, 'Game birds are decreasing—about 1/4 remain.' " From these remarks the statewide decline of 75 percent was derived [William T. Hornaday, *The Destruction of Our Birds and Mammals* (New York: New York Zoological Society, 1898)].

194. *Congressional Record*, 55th Cong., 3rd Sess. (January 13, 1899), pp. 628-31.

195. *Congressional Record* (February 24, 1899), p. 2303; (February 25, 1899), p. 2362.

196. The changes were suggested in a letter from Agriculture Secretary James Wilson to Lacey on January 17, 1900. Cart suggests that the revisions had been drafted by T. S. Palmer (Cart, "The Lacey Act," pp. 10-11).

197. See note 188, above.

198. *Congressional Record*, 56th Cong., 1st Sess., H. R. Report 474 (March 1, 1900), p. 2457.

199. Cited in Cart, "The Lacey Act," p. 11. Shields also mobilized the membership of the League of American Sportsmen through his journal *Recreation*.

200. *Congressional Record*, 56th Cong., 1st Sess. (April 30, 1900), pp. 4871-75; (May 18, 1900), p. 5704; 31 Stat. 187.

CONCLUSION

The fact must be emphasized that the history of conservation will show that a generation or more before that was made a principle to be applied to our other natural resources, the sportsmen of this country had established and applied it to the preservation of our game.

-Boone and Crockett Club, 1915

Institutions do not appear, nor do laws form, by some autonomous process detached from the interests of the politically powerful and abstracted from the historical processes that combine to form the present. Indeed, the view which emerges here is that public policy results from the often subtle interaction of interests within an institutional framework that is itself susceptible to modification by the very forces it constrains. The articulation of these interests arises out of individual and group responses to actual and prospective changes in well-being. Responses are constrained by law and custom, but these constraints are relaxed, strengthened, and otherwise modified both by deliberate action and by the interplay of social forces. Underlying these complexities is the real resource base that defines supply. But even supply is not wholly exogenous to the socioeconomy. Discovery, innovation, and technological change may extend or erode the known resource base; changes in values and tastes may define new resources or identify new scarcities. The growth in state regulation of wildlife during the nineteenth century illustrates well these intricacies of the public policy process.

This study has considered the process by which property in wildlife resources has become increasingly well-defined with scarcity. The particular path chosen has been shown to depend on wide-ranging considerations. Primary among these is the fugitive nature of the resource itself. Identifiable populations of most important species move not only across lines of private property but across state and national boundaries. The way in which this feature has been accommodated is a function of previous attitudes toward wildlife and of institutions developed to control access to its use, of the patterns of resource use that actually occurred, and of the response of governments to political pressures generated by perceptions of scarcity.

Before 1850, property rights in wildlife were poorly articulated. Although some states legislated restrictions on the reduction of animals to private property, the virtual absence of enforcement suggested unconstrained access. By 1900, wildlife was nominally protected by a large number of complex regulations that more precisely defined the nature of its common ownership. In effect, ownership in wildlife was assumed by the several states which, through the political process, determined the substance of these regulations. In no case did property in wildlife move in the direction of private ownership.

This study has focused on three interest groups—sportsmen, market hunters and game dealers, and landowners—each of which developed a demand for wildlife in an era of unconstrained resource use. Sportsmen were initially a small group, oriented toward the English view of field sports. The hunt was an aesthetic event and a healthful pursuit. Good sport could be found reasonably close to most major urban areas. The market for wildife was not large in 1850. The heyday of the beaver trade was past, though the demand for furs continued to sustain a force of trappers. The market for meat was undeveloped. In mild weather, difficulties in transportation and storage prevented the organization of a national market, but, even in winter, demand for game could, with some exceptions, be regionally satisfied. Whatever the volume of business, however, it was not constrained by law. Landowners and tenants were customarily subsistence hunters where game was abundant. Where noxious species threatened their crops or live-stock, they freely protected their property. The effects of land-use changes on wildlife, even where recognized, were not their concern. Early game laws often excepted the landowner from restrictions on hunting so that, even were they enforced, he was unconstrained in his behavior. Further, the hunters who entered private lands in search of game were few, so that the extent of damage sustained by the landowner was small.

In 1850, then, no group could expect to gain by investing resources in an effort to change the structure of property rights in wildlife in order to limit the access of others. Between 1850 and 1900, these groups ceased to operate in isolation. Demand grew, and the activities of each began to have significant effects on the activities of the others.

The number of sportsmen began to grow rapidly. While the wealthy remained the major spokesmen, with the increased leisure accompanying increased incomes, recreational hunters emerged from other income classes. Sport hunting, as opposed to hunting for the market, was a social event. Associations of hunters formed for the purpose of financing club houses, leas-ing shooting preserves, or simply fostering conviviality. Sev-eral popular sporting journals were able to coalesce their numbers as an interest group. Thus, sportsmen became a well-organized, homogeneous, and influential segment of the pop-ulation. They had among their numbers prominent business-men, politicians, and scientists.

The game market grew significantly during this period. The development of refrigerated railroad cars and cold storage increased the ability of dealers to contract for game from greater distances and market it throughout the year. Increas-ing urban populations, unable to hunt for themselves, pro-vided a growing demand for game. Sportsmen viewed the supplying of urban markets as the major cause of the decline in game populations. Through the public expression of the belief that "no wild species of bird, mammal, reptile, or fish can withstand exploitation for commercial purposes," they were able to secure legislation restricting the sale of game—first during periods when hunting was prohibited, and later year around. Similarly, in states that supported large wildlife populations, sportsmen were able to secure prohibitions on the exportation of game.

These prohibitions were facilitated by the formal assignment of ownership in wildlife to the state in which it occurred. This designation finds basis neither in English common law nor in the U.S. Constitution. It found early expression, however, in colonial charters that granted hunting and fishing rights to all citizens, and it was generally accepted by the states well in advance of the Supreme Court's dictum in 1896. The initial assumption of this power was no doubt by way of guarantee that European patterns of exclusive access to wildlife would not prevail in the "land of the free." The conflict between state and federal authority was not anticipated.

The strong historical bias in favor of unrestricted access to wildlife precluded the assignment of property in wildlife to landowners. Although this solution would have validated the sanctity of private property in land and would, further, have provided an efficient solution to the problem of allocating access to certain species, there was no basis for limiting the access of the "common man." At the same time, however, the increased numbers of hunters imposed real costs on landowners in the form of property damage. Thus, traditions of free access notwithstanding, landowners were accorded the right to protect their property from trespass. In the absence of restraints on the behavior of the landowner, however, protection from trespass was identical to assigning him property in wildlife. A measure of equality of access to the resource was maintained by subjecting landowners to the provisions of the game laws on an equal basis with all others.

By 1900, access to wildlife was severely constrained. The hunting seasons were becoming shorter, acceptable technologies fewer, and freedom in the disposition of the catch increasingly limited. States were beginning to assume the responsibility for protecting the wildlife populations that they had generally considered their property. Yet the most important constraint on access was the scarcity of game. Throughout most of this period, wildlife populations declined in the face of increasing protective legislation. Only in selected locations, where effective enforcement was coupled with favorable land-use patterns, were populations restored.

It has recently been argued that African ungulates might be harvested for their meat. The yield of meat per acre is believed to be larger than that available from domesticated species on the same land. Native species have a greater resistance to disease and are able to take selective advantage of the habitat. A similar case could be made in retrospect for the American bison; some even suggested it during the nineteenth century. But a national buffalo herd, stretching across the Great Plains, was irrelevant to the calculations of individual settlers. Their choice was between acquiring land for a farm or ranch and not acquiring it. In the face of thousands of private decisions (and an intensive hunt), the buffalo was eliminated. Similarly, a population of passenger pigeons of perhaps several billions might have provided a sustainable harvest of hundreds of millions annually. Yet the species required the hardwood forests that were cut over in exchange for the private income that lumber and cleared land could provide.

In the nineteenth century, there were effectively no constraints on land use. As the quality of the land changed, the distribution and abundance of wildlife changed. Those species which persisted were those able to colonize disturbed habitats and those requiring habitats not otherwise in demand. In the final analysis, the transformation of the landscape, perhaps the most important single cause of decline in the abundance and distribution of wild animals, entered only tangentially into the redefinition of property rights in wildlife.

Current conflicts over the control of wildlife and other natural resources are constrained by a mix of institutions and cultures at least as complex as that in evidence in this study. Conflict over wildlife resources has shifted to the international level. Mechanisms for international reconciliation are as undeveloped now as were parallel institutions at lower organizational levels in nineteenth-century America. That such mechanisms will further develop is very certain; that many species, most of them unidentified, will be extinguished in the interim is equally certain. Whether these are species which would have been saved were the mechanisms now in place, we will never know.

THE LOGIC OF INTERVENTION*

The collective behavior exhibited by sportsmen, market hunters and game dealers, and landowners is only partially explained by their own perception of the problems they face and the expected benefits of action taken to overcome them. Other aspects of this behavior are revealed by a more abstract view of the structure of these groups and the implications of this structure for collective action.[1]

THE SPORTSMAN

Under common property in wildlife, each sportsman could hunt whenever and wherever he chose, with his success dependent on technology, skill, and the local abundance of game. Other things equal, this abundance is inversely related to total hunting intensity, to which the individual sportsman contributes only a small part. At low hunting intensities, the opportunities of each hunter may be unaffected by the presence of others. At higher intensities, these other hunters may

*With apologies to Mancur Olson's *The Logic of Collective Action* (Cambridge, Mass.: Harvard University Press, 1965) from which this appendix draws more than its title.

impose costs on the sportsman by competing for existing animals and by reducing the ability of the population to replenish itself.[2]

As a result of these interdependencies, it is only in the extreme case where a single individual hunts an identifiable population that he will be able to reap the entire gain from a change in his behavior. Where the number of individuals is small, unilateral action may still be rational when the share of the gains to the individual is greater than the whole cost to him. Other hunters, then, would receive similar gains gratis. In the usual case of a large number of hunters, the change in abundance resulting from the partial abstinence of a single hunter will be so small that the share of the increase appropriable by the acting individual will not reward his action.

The problem becomes more complex in considering collective action. All relevant hunters might join together for the purpose of making voluntary changes in individual behavior which would yield collective benefits. Alternatively, a subset of hunters might attempt to force upon all hunters certain changes in behavior through a change in the law. Change resulting from the first form necessarily makes all hunters better off.[3] Change resulting from the second form may be favorable to all hunters or may redistribute income in favor of the membership of the organizing group.[4]

The benefits to be gained by a sole non-cooperator in a voluntary agreement are large, and the likelihood of the success of the first organizational form is small in the absence of a coercive force which, in effect, denies its voluntary character. This is not to imply, however, that several hunters together would not gain by restricting their behavior in the absence of cooperation from others. First, their share of gains might be sufficiently large to merit a change in behavior. Second, they might be willing to sustain costs in the present with the expectation that their example would lead to the general modification of the attitudes and behavior of nonmembers, thus providing future benefits. Third, they may derive satisfaction from the mere fact of their own abstinence if they feel this behavior to be particularly moral or righteous. Finally, it should be noted that a subset of hunters may engage in market transactions in order to secure improved access to wildlife through the purchase or rental of shooting lands. The group must weigh the market power to be gained by the increase in membership with the costs to existing members of diluting potential gains.

The utility of these strategies depends on the costs and benefits associated with their implementation. Viewed in the narrowest sense, in which wildlife is useful to sportsmen only as an object of the hunt, the benefits of action are equivalent to the costs that would be incurred as a result of inaction. By the economist's measures of consumers' surplus, costs of scarcity depend on the individual's appraisal of the substitutes that exist for sport hunting. The better the substitutes, the smaller the benefits from actions which seek to mitigate scarcity. Those who considered themselves true sportsmen did not depend on killed game for food. To the extent that a particular species was considered a toothsome treat, it could have been purchased in the market, and probably at a smaller dollar cost including the amortized expenses for guns, ammunition, clothing, and other accoutrements of the hunt. That there must have been benefits derived from the sporting experience is indicated by the fact that market purchases were not a substitute for the hunt itself.[5] It was never clear that the utility that the sportsman gained from the hunt increased with the quantity of game killed. In fact, though it was partly an adaptation to the conditions of scarcity, many sportsmen found virtue in "hunting" by camera as this became technologically feasible, and others altered their conception of "game" to fit the relative distribution of species.[6] Finally, there were a number of activities which the sportsman might have found acceptable as substitute for the hunt. Some were closely related, as were trapshooting and riflery, and others, such as archery, yachting, and equestrian sports, were similar only in general social style.

Against these benefits, sportsmen must weigh the costs of collective action. Organization, at least within a city or small region, was relatively costless. The individuals concerned were largely of the same social class and likely had social or business contacts previous to any consideration of organization over sporting matters. Even these costs would have been minimized to the extent that a group of sportsmen had already organized for the purpose of securing a hunting lodge or a shooting preserve.

Once the group is organized, there are additional costs in taking action. A group seeking to acquire access to wildlife on private lands must first determine a course of action among its members. The larger the membership, the greater the command over resources but also the greater the difficulty in reaching internal agreement. In addition, where a well-developed market does not exist, there are costs associated with bargaining between landowners and sportsmen, particularly, as was common, when a number of contiguous holdings were leased in a single package. The second type of contractual costs are those incurred in attempting to direct the course of legislation. These range from the concrete costs of letter writing and travel to the more imprecise costs of future political support or the exercise of personal influence.

Finally, there are enforcement costs. Where a market transaction or voluntary agreement is made by contract, enforcement costs have been included to the extent that this type of agreement is backed by the existing laws of private property and contract. The costs of ascertaining whether or not the terms of the contract are being met may, however, be significant. For legislative enactments, the distribution of costs is assigned by the law itself. Where these costs are assigned to a group other than that seeking the passage of the law, as for example, to a government agency, contractual costs may rise by reducing the likelihood of legislative success. Before 1885, most game legislation provided for enforcement by the division of levied fines between the private prosecutor and the state or local government. Against these gains, however, the individual had to set the possible loss of court costs in an unsuccessful suit. Sportsmen's groups often saw it in their interest to enforce laws of this type in order to secure some of the benefits of the legislation. Such groups, in fact, provided the only law enforcement prior to the creation of state-level wardens.

In spite of the small organizational costs, the small likelihood of capturing a substantial portion of benefits as well as the small size of those benefits suggest only marginal incentive for organization by sportsmen. Additional incentive, however, derived from more nebulous considerations described in Chapter 2. The social role of the conservationist in an age of rapid resource use is not to be underestimated, and by recommending the conservation of a resource on which no member of his class depended for his fortunes, the gentleman sportsman could play that role without risking the loss of social or economic status.

THE MARKET HUNTER

The selective arrangements that develop and which persist among market hunters must in some manner be related to the external costs which would be generated in their absence.[7] Where an individual hunter's activities in no way affect other hunters, externalities are nonexistent, even in the absence of structured property rights. Conversely, where interactions are frequent and complicated, externalities do not present a difficulty if property rights are adequately defined. Where interactions are few, the extensive definition of property rights, although mitigating limited external costs, requires a large expenditure of resources in devising and enforcing the chosen structure of rights. Where interactions are many, loosely defined property rights, although calling for reduced expenditures in these ways, result in the reduced efficiency accompanying large external costs. One would expect that, other things equal, the greater the frequency and intensity of social interaction, the more structured that interaction will be. Informal codes of behavior were generally more prevalent near waterfowl feeding grounds than, for example, in the Maine woods. Before state laws effectively regulated the behavior of waterfowl hunters in the Currituck Sound region of North Carolina, they had achieved a quite well respected understanding that limited shooting to Mondays, Wednesdays, and Fridays.

Market hunters, in addition to engaging in collective activity to modify behavior among themselves, might also organize for the purpose of providing a common front against other groups who sought access to wildlife. If hunters supplying a particular species to a given market could monopolize its capture, total industry profits must increase for any given demand. Data are sparse, but the substitutability among species of game as well as between wild species and their domestic counterparts suggests that significant price increases would be met by exit from the market. Supporting this claim is the fact that the several urban markets were separated only by transportation costs, particularly as refrigerated storage and freight became feasible, so that the price in one market could not rise above the supply price, including transportation, of other markets.

Against these benefits, which appear small, costs appear large. The monopolist must gain exclusive control over all habitats for the species in question in the range serving the market to be controlled. This involves bargaining with each landowner to guarantee access for monopoly hunters and to prohibit access for other hunters. Bargaining with existing retailers would be insufficient, first because it would allow subsistence hunting and, second, because it would provide great incentive for a new retailer to enter the industry to deal with nonmonopoly hunters. An additional complication is that monopoly hunters, in attempting to regulate their internal affairs, may make their activities more susceptible to control. Hunters working for themselves may be able to engage in illegal hunting with less risk, and, to that extent, the desirability of formal organization, at least from the viewpoint of the individual hunter, is reduced.

The likelihood of this kind of collusive behavior seems small. Consider instead organization for the purpose of influencing the legislative process. The potential benefits relate, as with sportsmen, to the costs of increased scarcity of wildlife. With a fixed market price, the individual hunter would prefer a greater to a lesser density of game since a larger quantity could be profitably captured. Increased density is achieved, however, only by the reduction of hunting intensity in the present. But even if an increase in future density was both desired and achieved, the individual hunter would not necessarily be able to capture his share of the gains since additional hunters would be drawn into the industry by decreased capture costs.

Whatever the gains from changes in legislation, the costs of organization would be large. Market hunters, particularly in upland regions, were widely dispersed, and securing support from each would be costly even if all were strongly in favor of a given program. The additional difficulties of the free rider and of the uncertainty of the legislative process make these costs even higher. The enforcement costs would likewise be enormous since there is no reason to expect voluntary compliance with restrictions. If all hunters complied, the individual hunter would gain by ignoring the law, and if all other hunters did not, he would lose by compliance. Thus, compliance would only follow from extensive enforcement undertaken by independent agents and accompanied by stiff penalties for illegal behavior. Sportsmen were able to abide by the game laws even in the absence of enforcement by the force of the ethic of fair chase for which the market hunter had little use.

THE LANDOWNER

An individual who owns a parcel of land owns, in effect, a bundle of property rights. The value of the land depends critically on the composition and stability of that bundle of rights. When that bundle is stable, the property can be used without ambiguity. With the likelihood of change, the owner stands to suffer losses or reap gains to the extent that the bundle is reduced or augmented. It should be clear that it is in the interest of the landowner to seek institutional change that yields windfall gains, protect himself against change that yields windfall losses, while at the same time balancing the frequency of change against the costs of uncertainty. Important rights that the landowner might seek to gain or to protect include the right to alter landscape, the right to kill or preserve wildlife existing on his property, and the right to regulate access to his property. Controversies over these rights summarize the role of the landowner in the conflicts over the allocation of wildlife resources during the period 1850-1900.

The strategic implications of these circumstances may be pursued at several organizational levels. To the degree that any landowner has control over wildlife existing on his land, he may lease or sell access unilaterally. For example, he may lease exclusive access to a group of sportsmen, he may open his lands to all comers on a fee-for-use basis, or he may levy a charge per animal removed. These solutions apply equally well to market hunters or, in the case of the first, to those who would create a sanctuary in which hunting is prohibited. The value of any such contract is heavily dependent on the degree of control which the landowner has over wildlife existing on his land. If the contract includes the right to hunt any species at any time of the year, using any method, it is more valuable than if any or all of these dimensions are circumscribed. During the period 1850-1900, the contract might have been limited in value by legislation stipulating the seasons in which particular species could be hunted, the methods that could be employed, the need to procure a hunting license, bag limits on some species, and constraints on both the exportation of game and its sale.

In two-party agreements of this kind, transaction costs are relatively small. The contract could be drawn up in bargaining between a single landowner and a single individual desiring access for himself or for a group of hunters. The policing costs are another matter. In the case of an exclusive agreement between a landowner and a sportsman's group, each has an interest in excluding other hunters from the land—the landowner because of additional damage for which he would be uncompensated and the sportsmen because other hunters reduce the availability of game. Protection of the interests of one protects likewise the interests of the other, so that the question of responsibility for enforcement is a matter for bargaining. Both parties would benefit by third party enforcement, and as states began to take an active interest in wildlife protection, public agents were available for such duties.

Contracts involving a single landowner are unlikely where the parcel is small relative to the habitat of an identifiable population in demand. Whereas a fifty acre parcel may have market value as a quail-hunting ground, its value for deer hunting is minimal. On the other hand, for species which have regular and dense patterns of migration, a location along the route may have great value in spite of its small size. Likewise, a site adjacent to a waterfowl feeding ground may be similarly valuable. But for species that vary widely over a locale and/or which occur in low densities, the hunter will want the freedom to roam accordingly. Although the individual (or club) may bargain separately with a number of landowners in order to secure access to a sufficiently large area, he may expect to retain the upper hand only to the extent that each landowner remains ignorant of actual or desired bargains with his neighbors. This seems unlikely, however, and instead, the sportsman may find a strategically located landowner holding out for a larger share of the gains. Alternatively, landowners in a region may market collectively access to their lands. In this case, costs are increased by the need for agreement on the conditions of the offer and on the internal division of receipts. To the extent, however, that the relevant group of landowners is already organized, as for example, in a local agricultural society, these costs are reduced. Conversely, the organization may discover that once it has organized for the purpose of negotiating access to wildlife, it may engage in other kinds of collective action with smaller costs.

Neither individual nor collective market transactions are useful in altering access to wildlife to the extent that ownership lies with the state. As has been indicated, there are public policy changes that the landowner would view as desirable and other changes that he would seek to prevent. The conditions under which landowners would seek to influence public policy are analogous to those for sportsmen and market hunters and need not be elaborated here.

NOTES

1. For example, Reiger observes that "wildlife preceded forests as the most important environmental issue." Where, he asks, were the forestry clubs and forestry journals comparable to those concerning wildlife [*American Sportsmen and the Origins of Conservation* (New York: Winchester Press, 1975), pp. 51-52]? The answer in part is that mem-

bers of local forestry clubs, by the very nature of the resource with which they would have been concerned, stood to gain little whatever their knowledge and preferences.

2. Additional costs may be imposed by "crowding," which increases the probability that a hunter will frighten away another hunter's deer, or be mistaken for one.

3. This assumes that each member correctly anticipates the implications of the decision for himself.

4. The success of the group might depend on the extent to which its goal speaks to the general welfare.

5. Consumer's surplus understates the gains accruing to the hunter-consumer by the amount of the producer's surplus. If market price is determined by the cost of capturing the marginal unit sold, the sportsman who hunts in an area of relative plenty, where capture costs are below market price, retains this additional difference.

6. See the discussion of the changing attitudes of the sportsman in the face of scarcity of traditional game species, Chapter 3, text at notes 20-29.

7. See Harold Demsetz, "Toward a Theory of Property Rights," *American Economic Review* 62 (May 1967), 347-59, and Garrett Hardin, "The Tragedy of the Commons," *Science* 162 (1968), 1243-48.

APPENDIX 2

AN ECONOMIC MODEL OF WILDLIFE HARVEST

This book has, throughout, relied on narrative argument. The scope of the analysis and the available data dictate this approach. Nonetheless, some readers may wish to consider these historical dynamics within a more rigorous analytical framework.[1] This appendix is meant to sketch the outline of such a framework.

The regulation of access to wildlife is historically based on the protection of species during the breeding season. The model developed here assumes the initial existence of such protection, and it is built around the relationship between the size of the harvest during the hunting season and the ability of the population to regenerate itself during the breeding season. This treatment allows for the consideration of policies which affect the relationship between harvest and reproduction as well as for the development of a more realistic model of population growth than is often assumed in the literature. Further, the model considers the economic sector in terms of supply and demand functions. Although this approach obscures the precise distinction between changes on the extensive and intensive margins of production, it allows for the identification of the market-clearing harvest in each period as a function of the stock of wildlife. This provides the basis for the description of the regulatory measures employed during the period 1850-1900 in terms of their effects on the size of the harvest, the economic surplus generated, and the wild stock remaining. It therefore provides the basis for evaluating these measures from the viewpoints of the several interest groups concerned with the allocation of access to wildlife. This is an essential feature in the analysis of historical events that appear to depend critically on the perceptions of economic and political agents.

THE ECONOMIC SECTOR

The basic features of the economic sector of the model are indicated in Figure 1. There are two quantities that must be indicated for any given time period—the size of the wild population and the size of the harvest. The size of the wild population is measured left from the price axis. (The merits of this unconventional formulation will be evident below.) The size of the harvest can only be indicated on the figure once the size of the wild stock is known. Supply of captured animals is assumed to be a function of the cost of capture, and the shape of the schedule is largely attributed to the increased effort required as the species becomes scarce. Two assumptions underlie this relationship: first, where habitat is uniform, the greater the density of the species, the greater the likelihood of finding an individual animal in any particular location; second, where habitat is varied, animals will tend to select the more secure location first, and only with increased numbers will more vulnerable locations be populated. It is further assumed that at some small population, N_2, the marginal cost of capture increases at a faster rate and that at some large population, N_7, the marginal cost of capture ceases to decline (or declines at a slower rate).[2]

The initial assumption about the demand for the species in question is that it is a function only of the quantity of wildlife captured. Its shape and height are thus unrelated to the size of the wild population. The curve need only be located according

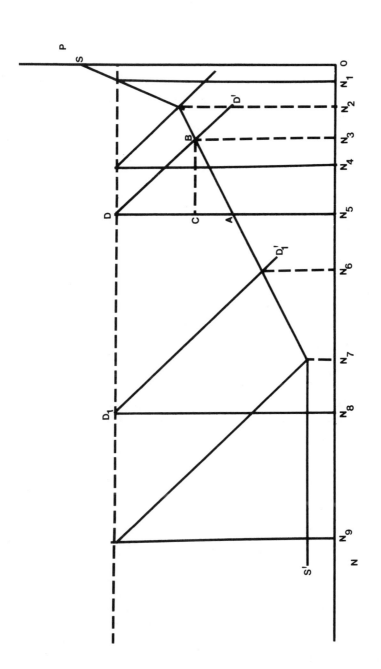

Figure 1

267

to the stock of animals at a given time. As the stock changes, the curve makes parallel shifts with an unchanged intercept on the moving price axis.

Assume, in Figure 1, that ON_5 is the size of the wild population at the beginning of the open season, DD' is the demand for the species, and segment AS of the overall cost of capture curve SS' is the supply schedule. The equilibrium harvest will be N_5N_3, leaving a breeding stock of ON_3. Additional captures are characterized by marginal cost greater than marginal benefit and are therefore not made. Total surplus derived by hunters who consume their own catch is ABD. If the harvest is marketed, price will be BN_3, and consumer surplus BCD accrues to final purchasers.

Given the structure of supply and demand, the market-clearing harvest consistent with the initial wildlife stock ON_5 can be plotted as point A in Figure 2, where the distance AN_5 equals the distance N_5N_3 in Figure 1. A similar point in wildlife stock-harvest space can be plotted for each initial stock. For example, stock ON_8, for which the relevant demand curve is D_1D_1', yields a harvest of N_8N_6. This is plotted as point B in Figure 2, where BN_8 equals N_8N_6. The equilibrium harvest for any initial stock between N_4 and N_9 lies on a line passing through points A and B. For stocks larger than N_9, the harvest is constant at N_9N_7. For stocks between N_4 and N_1, the slope of the harvest curve rises to reflect the greater rate at which capture costs rise. Harvest falls to zero at N_1, where the maximum demand price equals the marginal capture cost. The location of the intercept of the harvest curve with the axes in Figure 2 depends on the relationship between maximum demand and supply prices. If maximum demand price is less than maximum supply price, as in Figure 1, there is a stock below which the population will not be reduced by harvest. Conversely, where maximum demand price is greater than maximum supply price, there is a range of stocks unable to satisfy equilibrium harvest, and which therefore lead to extinction. Thus, the curve must intersect the vertical axis. Finally, where maximum demand and supply prices are equal, the curve emanates from the origin.

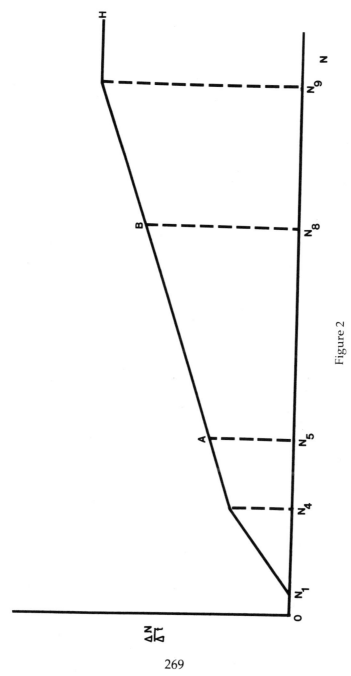

Figure 2

THE BIOLOGIC SECTOR

In order to determine the size to which the population will be restored following the harvest, some assumptions about the dynamics of populations must be made. The model to be initially assumed is the discrete analog of the form often employed in the literature.[3] That model takes the rate of growth of population, dN/dt, to be of the quadratic form $rN + cN^2$, where r = the natural rate of increase, and $c = -r/K$, where K is defined as the carrying capacity or maximum population which a habitat can sustain. Thus, $dN/dt = rN(K-N/K)$. Figure 3 shows the rate of change of N as a function of N itself. This figure recognizes the possibility of extinction by assuming $\underline{N}$, the smallest population at which reproduction occurs, to lie at some positive N. Finite populations may fail to reproduce owing to the difficulty of finding mates or, in colonial species, the inability of small numbers to trigger reproductive behavior.[4]

Analogously, it is assumed here that $\Delta N/\Delta t$ is a function both of reproductive potential and of density dependent constraints. A relationship between $\Delta N/\Delta t$ and N may be hypothesized as in Figure 4. The height of the curve for a given N indicates the net increment to the population during one cycle. Thus a population N_t at the beginning of one breeding season gains the number of individuals equal to the distance N_tN_t'. At the beginning of the next season, it contains $N_t+N_tN_t' = N_{t+1}$ individuals. During the second period, the population gains $N_{t+1}N_{t+1}'$ and, at the beginning of the following breeding season, contains $N_{t+1}+N_{t+1}N_{t+1}' = N_{t+2}$ individuals. In successive periods, the population increases to N_{t+3}, N_{t+4}, etc., until the stable population $\bar{N}$ is reached.

THE DYNAMICS OF THE HARVEST

By combining Figures 2 and 4, we can locate the dynamic equilibria between harvest and reproduction, if any exist, and determine their stability. Because we are assuming discrete rather than instantaneous responses, the intersections of the curves are not themselves the locations of equilibria, though they may indicate the existence of nearby equilibria. Several types of relationships are illustrated in Figures 5-9.

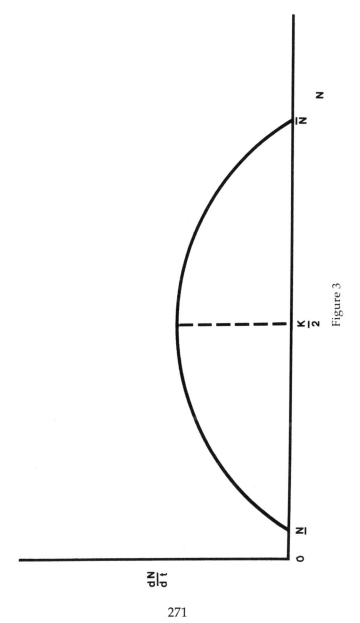

Figure 3

271

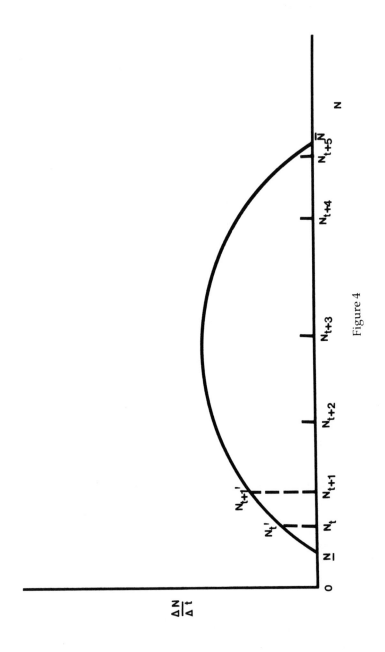

Figure 4

272

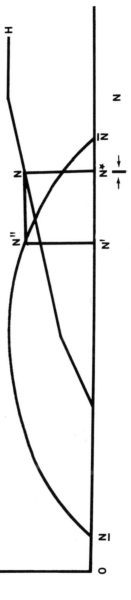

Figure 5

In Figure 5, there is one equilibrium at N^*. An initial population of N^* induces a harvest of N^*N during open season, which reduces the population to N' (where $N^*N = N^*N'$). Population N' is increased during the breeding period by $N'N''$, which restores the population to N^* (where $N'N'' = N'N^*$). Furthermore, N^* is stable. Any initial pre-harvest population greater than N^* is characterized by harvest greater than growth, and any below N^* by growth greater than harvest. Figure 6 shows two equilibria. N^* is stable as in Figure 5, and N^{**} is unstable, which may be analogously demonstrated. Any population smaller than N^{**} leads to extinction. Figure 7 illustrates the role of the increased slope of the supply curve for small populations in reducing the likelihood of extinction. Without this feature, the harvest curve would continue along the dotted line, and extinction would follow from any population smaller than N^{**} as in Figure 6. The increased slope provides a third, stable, equilibrium at N^{***}.

In addition to these general cases, there are two polar cases —one in which equilibrium harvest is so large that additions to the population are unable to compensate for harvests at any population size, and the other, where no harvest occurs even at $\bar{N}$. These are illustrated in Figures 8 and 9. It should be noted that the case illustrated in Figure 8 does not require that the harvest curve lie entirely above the production curve. It may intersect that curve such that its height is always greater than the horizontal distance between the two curves.

It is a primary aim of this model to aid in the understanding of the effects of alternative regulatory measures on the equilibrium stock of wildlife and, in turn, on the welfare of interest groups concerned with the distribution of access to the resource. The measure of welfare may be taken as the sum of producers' and consumers' surplus. For a single time period, the surplus has been defined in Figure 1. It should be noted, on the one hand, that the size of the surplus increases with the size of the wild population due to the declining supply curve, but, on the other hand, that for any given stock, the size of the surplus increases with the size of the harvest. Since policies which tend to increase equilibrium stock also tend to decrease surplus consistent with any given stock, there may not exist, in

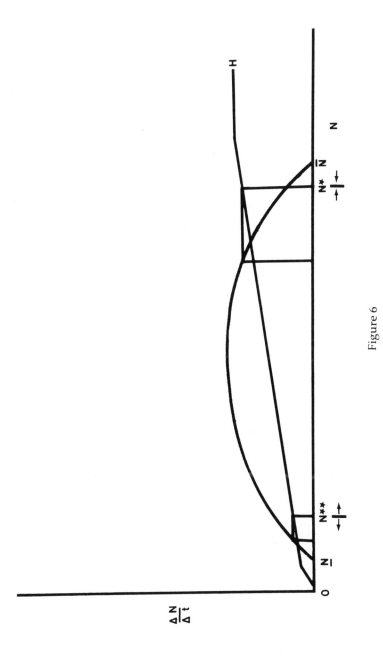

Figure 6

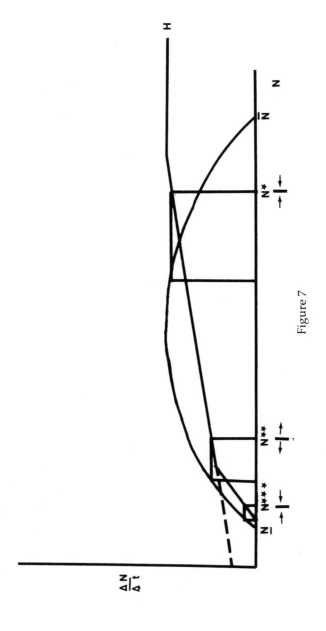

Figure 7

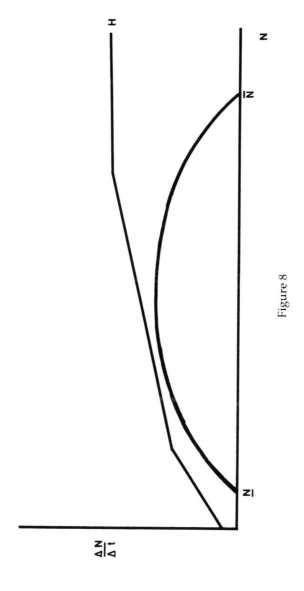

Figure 8

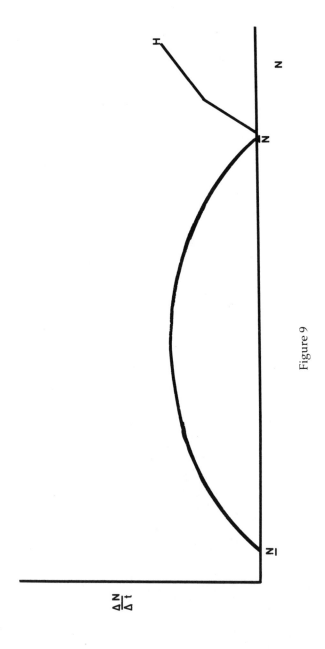

Figure 9

278

any particular example, policies which are unambiguously preferable in that they increase surplus in all time periods.

Consider the case illustrated in Figures 10 and 11. The initial supply curve, S_1, yields a harvest schedule H_1, which shows a stable equilibrium at stock ON_1. Assuming that equilibrium has been reached, an annual surplus of ABC is being received. Suppose a tax is imposed that raises the supply curve to S_2. The harvest schedule consistent with this change is H_2, and the new equilibrium is found at stock ON_2. Once this new equilibrium is reached, surplus FGH, greater than ABC, will be gained. However, the path to this equilibrium is marked by harvests with surpluses smaller than ABC, beginning with surplus ADE at the initial equilibrium ON_1.[5] The evaluation of the policy which imposes tax S_1S_2 can be made by comparing the present values of the surplus streams resulting from harvests under curves S_1 and S_2. In theory, the optimal path can be identified and the corresponding policy implemented.

Nineteenth-century wildlife was not managed with any explicit concern for maximizing the present value of benefit streams. Indeed, in these formative years, wildlife was hardly managed at all. Policy was created as interest groups were able to organize successfully for political purposes. All groups spoke to the greater good they represented: sportsmen, that future generations of sportsmen might benefit from the hunt (and the rest of society from their leadership, improved as a result); market hunters, that the legitimate demands of the hungry, cold, and fashion conscious might be met; naturalists, that future generations might enjoy and benefit from a rich natural world; humanitarians, that animals might be spared suffering and death; and landowners, that the rights of private property might provide the strongest base for the nation's future. Each policy supported some visions of the greater good and denied others. At least two sources of differential support for regulatory proposals can be discerned. The first was that different rates of time preference could not be reconciled. Sportsmen often claimed to place great weight on the stock of wildlife that would be available to future generations, whereas market hunters often appeared to limit their concern to several

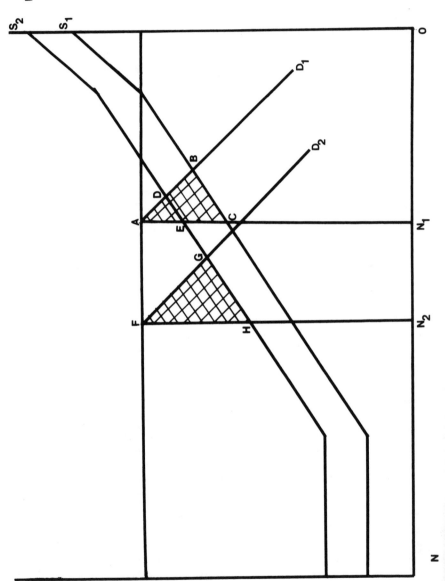

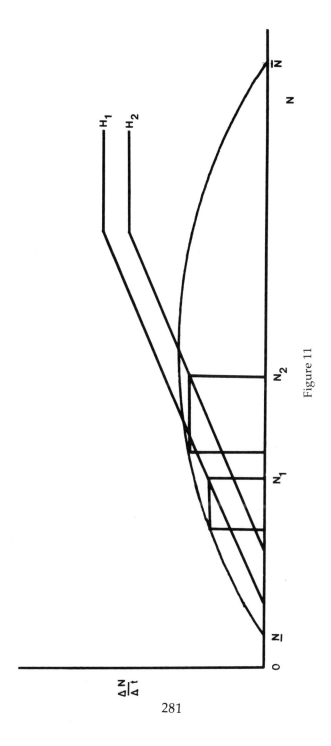

Figure 11

281

years. Part of this difference may have been due to divergent estimates of the size of present populations or of the impact of harvest on future populations. Market hunters, for example, may have overestimated abundance and underestimated the impact of harvest.

More important than these intertemporal considerations were those which altered the distribution of gains from hunting among hunters in a given time period. Suppose that the tax which shifted supply from S_1 to S_2 in Figure 10 was in the form of a limitation of hunting technologies. Since sportsmen and market hunters often employed different technologies, the loss in economic surplus occasioned by the limitation of methods favored by the market hunter would be born largely by that group, both in the present and in the future. The class of sportsmen, on the other hand, would gain in all periods. Irrespective of the present value of surplus resulting from this change, it would be expected that sportsmen would favor it and market hunters oppose it. A similar alignment of interests might be expected over measures to restrict the marketing of game.

THE EFFECTS OF INSTITUTIONAL CHANGE

Having explored the dynamics of the model, we turn to the consideration of various taste, technological, and institutional changes which occurred during the period 1850-1900.

It has been assumed to this point that the height and shape of the demand curve are fixed for a given species. It might be argued, however, that the character of demand is a function of the size of the wild poulation. The case can be made for a shift in either direction, and while this gives ambiguous results for the overall model, it would be feasible to determine, for the historical case, the direction of the shift by interest group and species.

The rarity of a species makes its capture a valued prize, and hunters may thus be willing to expend greater resources for its capture than were it common.[6] The demand curve shifts upward as it moves to the right. Thus, the harvest in any given period is larger than it would be in the absence of such induced changes in demand, though the increase is greater the smaller the wild stock. In terms of the above figures, the consideration of such forces increases the likelihood of an unstable equilibrium as in Figure 6, and consequently introduces the risk of extinction where it may not have existed previously.

The other induced shift in demand derives from the social costs of scarcity that hunters impute to their own behavior. These include the increased cost of capturing scarce animals, the lost reproductive potential of captured animals, the lost aesthetic utility to nonhunters, and, as extinction approaches, the loss of the species' gene pool. Demand would fall as the wild population decreases, and the harvest schedule would shift rightward and increase in slope. The implications for the model are the reverse of those for the previous case—a reduction in the likelihood of extinction and an increase in the equilibrium population.

Aside from these induced shifts in demand, there was a general growth in demand for wildlife during the period 1850-1900, both as marketable game and as the sportsman's target. This factor alone indicates for demanded species the reduction in the size of stable stocks or even their extinction. In addition, some species, such as the beaver and certain birds, faced fluctuating demands according to the whims of fashion.

Changing technologies, which affect costs, have similar effects on the quantities taken in any period and consequently on the equilibrium stock. The period 1850-1900 was characterized by falling costs, both with improved transportation in bringing game to market and in taking hunters to game. Refrigerated railroad cars and cold storage drastically reduced the costs of delivering game to market in temperate seasons. Finally, the development of the breech-loader, punt and swivel guns, and later, of automatic weapons, facilitated the capture of game and consequently lowered costs. All of these changes increased the size of the harvest in a given time period and, in the same manner as increases in demand, tended to reduce the stock of the target species.

It remains to describe the effects of regulatory changes on the dynamic process. Consider first the length of the season. With a shorter season, the time during which the hunt may be conducted is reduced, and with less freedom, the cost of making the hunt will rise, shifting the supply curve upward. The character of the shift depends on the extent to which the relatively more profitable times have been excluded from the season. In addition, the increased density of hunters that may result from restricted seasons imposes such congestion costs as the increased likelihood of being injured or killed, of prospective targets being scared away or captured, and of resources lost in arbitration over disputed game.[7] The extension of the reduced open season is the closed term, in which hunting is prohibited for a period of years. This effectively raises the supply schedule to the point where no harvest is taken, and the population is allowed to grow as indicated by the population response curve for the duration of the restriction.[8]

Another type of regulation which shifts the supply schedule is the limitation of hunting technologies. As the abundance of a species changes, the profitability of various methods of capture is presumed to change. The effect on the supply curve is dependent on the particular technologies excluded. Punt and swivel guns, firing huge charges, would be efficient only where the likelihood of killing a large number of waterfowl with one shot was great. The prohibition of such weapons would not affect that portion of the supply curve where the most efficient technology is a smaller gauge shotgun. The supply curve might twist or tilt rather than shift upward. Some changes in technology would have more complex effects. Hounding deer, for example, makes the animals wary, and hunting them becomes more difficult for those employing other methods. The restriction of hounding would, with some time lag, lower costs for those preferring other methods. In a similar way, the ruffed grouse, initially so unwary of man that they could be knocked from trees with sticks, have become very difficult targets. The supply curve has thus risen over time.[9]

A REEXAMINATION OF WILDLIFE POPULATIONS

We must now turn to a more detailed exploration of the behavior of wildlife populations.[10] The way in which the population response curve has been drawn in all above cases was carefully chosen so that the growth path from any initial N would converge on Ñ, the carrying capacity. Consider, on the other hand, Figure 12. If the population is at N at the beginning of the breeding period, then $\Delta N/\Delta t = NN'$, indicating that the new stock, $N + NN' = N^*$, is greater than Ñ. This is not a result to be avoided. The population response curves assumed to this point have neglected the problem of the age structure of populations. A population which has achieved long-run equilibrium with its habitat has, by definition, a constant age distribution. During the early stages of growth, however, the young will compose a disproportionate share of the population. Their environmental demands are generally less than those of adults. At some particular stage of growth, the number of existing animals would, were they proportioned as in the equilibrium populations, equal the carrying capacity. Births at that time, however, take into account only the present relationship of the population to carrying capacity, and not the future relationship of the present population. Consequently, more young are born and survive than can be sustained in the long run. As this population matures, carrying capacity will be exceeded. By definition, this situation cannot persist.

The adjustment process depends on the responsiveness of the population in question. It may take the form of a dampened oscillation that approaches carrying capacity in the long run, it may fluctuate between populations above and below the carrying capacity, or it may even take the form of an explosive oscillation that is limited by the convexity of the population response curve between N and Ñ.[11] Combined with any of these adjustment paths may be an actual reduction of carrying capacity, temporary or permanent, as the result of habitat damage accompanying overpopulation. Excess browsing, for example, may kill vegetation or reduce the production of foliage in a future growing period.[12]

These features may easily be included in the model by allowing $\Delta N/\Delta t$ to assume negative values for N's greater than Ñ. This adjustment will be modified in consideration of harvest activity. Only in the extreme case will the net change in population greater than Ñ be the sum of the harvest and the decrement indicated by the vertical distance between the

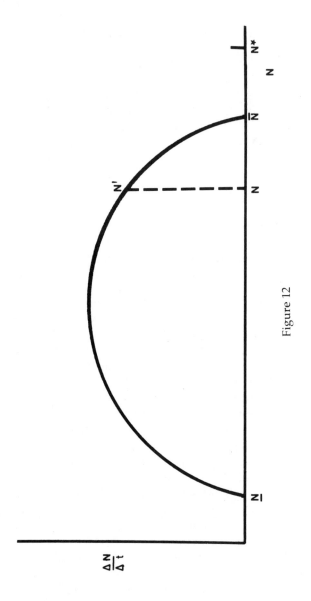

Figure 12

population response curve and the horizontal axis. To the extent that the harvest reduces pressure on resources by reducing population, the decrement which the population would suffer were it adjusting in the absence of human harvest is reduced.

The general issue of the relationship between harvest and natural decrements must be considered more fully. To this point, the difficulties of intra-year changes in population have been ignored. The population response curve has implicitly assumed that changes in $\Delta N/\Delta t$ are fully accounted for by changes in births. It is surely not true, however, that a population that is approaching carrying capacity has a birthrate which, in itself, is responsible for a near zero increment in population. Rather, at this density, decrease and increase are nearly equal, leaving a small net change. If the birth rate is assumed to be constant, then actual births are represented by a straight line with slope r, which must lie wholly above the population response curve, as shown in Figure 13. The literature suggests, however, that r itself changes as N changes, and although it may never fall to zero within relevant ranges of N, its decrease as N rises gives a convex shape to the response curve (curve RN in Figure 13). At population N^*, N^*N'' are born, while $N'N''$ of the total, $N^* + N^*N''$, fail to survive until the next breeding period.[13] This is the result which the initial model provided. The model assumed, however, that $N^* + N^*N'$ made up the entire population from which the harvest might be drawn. It is seen, however, that not only N^*N' but also $N'N''$ might be harvested without reducing the population below N^*, because those individuals would not survive in any event.[14]

The applicability of this modification to a situation of human harvest is not altogether straightforward. The animals that compose $N'N''$ are surely not a random sample of the population. In fact, they consist mainly of the young, the old, and the sick. Natural predators might be expected to select prey from this group, but there is no reason to expect human hunters to do so. Rather, those deliberately hunted tend to include the most fit and able members of the population. Not only might this behavior lower the growth response of the population by selecting those which have the greatest reproductive value,

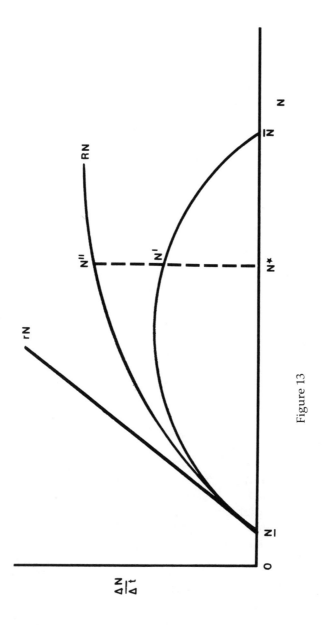

Figure 13

288

but it may also affect the "quality" of the population in the future by working counter to evolutionary forces. This implies that the relevant reproduction curve lies below RN, the true curve, in Figure 13, and its precise location varies according to the selectivity of the human hunt.[15] More important, this suggests that the effective curve can be shifted by regulation altering hunters' targets. Modifications in the deer season are often argued on these grounds. A policy that encourages hunters to include in their bags more of those individuals that would otherwise die between one breeding period and the next would increase stable equilibrium from N^* to N^{**} in Figure 14.

This model may be further modified to consider the complexities which derive from the interrelationships among wildlife populations. Changes in these interrelationships imply changes in the focus population. For example, the removal of a predator that kept two or more prey species in competitive equilibrium may allow one species to dominate. Thus, the removal of predators in Yellowstone Park has allowed the elk largely to replace the white-tailed deer and beaver.[16] Similarly, the introduction of exotic species may reduce the equilibrium size of native populations. These effects can be included in the model by shifts in the population response curve to indicate changes in carrying capacity and/or in reproductive behavior.

Finally, the model may incorporate the effects of land use changes which alter both the quantity and quality of habitat. During the period of this study, changes in the distribution and abundance of wildlife which occurred as the result of habitat alteration were largely fortuitous. Habitat improvement, winter feeding, and other management strategies were only sparsely employed. They may, nonetheless, be evaluated within the framework of this model.

NOTES

1. The model presented in this Appendix is an expository device and is, therefore, neither a critique nor an elaboration of the growing body of literature on the exploitation of wildlife populations. The interested reader is referred to Vernon L. Smith, "Economics of Production from Natural Resources," *American Economic Review* 58 (June

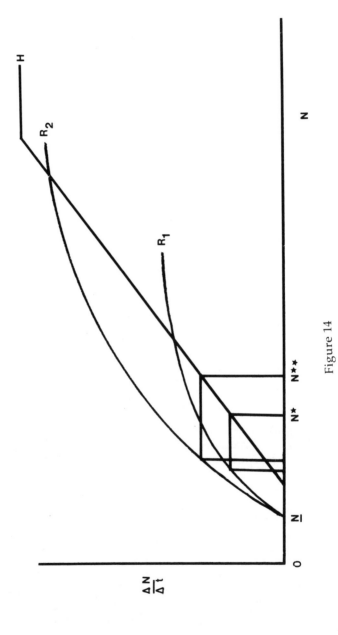

Figure 14

1968), 409-31; Vernon L. Smith, "On Models of Commercial Fishing," *Journal of Political Economy* 77 (March/April 1969), 181-98; H. Scott Gordon, "The Economic Theory of a Common Property Resource: The Fishery," *Journal of Political Economy* 62 (April 1954), 124-42; Colin W. Clark, "Profit Maximization and the Extinction of Animal Populations," *Journal of Political Economy* 81 (1973), 950-61; Colin W. Clark, *Mathematical Bioeconomics: The Optimal Management of Renewable Resources* (New York: John Wiley & Sons, 1976); Richard S. Miller and Daniel B. Botkin, "Endangered Species: Models and Predictions, *American Scientist* 62 (March-April 1974), 172-81; Jon R. Miller and Frederic C. Menz, "Some Economic Considerations for Wildlife Preservation," *Southern Economic Journal* 45 (January 1979), 718-29; and Frederick M. Peterson and Anthony C. Fisher, "The Exploitation of Extractive Resources: A Survey," *Economic Journal* 87 (December 1977), 681-721.

2. A more general, non-linear, supply schedule might be drawn convex to the zero-stock origin, thus eliminating the need to identify points N_2 and N_7, but the complexity of the model would be considerably increased without a corresponding increase in expository value. The particular shape of the supply schedule would depend on the characteristics of the species in question. Deer and upland game birds, dispersed fairly evenly over their ranges, might be characterized by the supply schedule shown in Figure 1. The supply schedule for shore birds and waterfowl may have an extended flat portion, rising at the point where feeding or nesting areas are disrupted or depleted.

3. See, for example, Lawrence Slobodkin, *Growth and Regulation of Animal Populations* (New York: Holt, Rinehart and Winston, 1961), pp. 60-61; and Ian A. McLaren, Introduction, in McLaren, ed., *Natural Regulation of Animal Populations* (New York: Atherton Press, 1971), pp. 1-6. The discrete form is employed by Dasmann, *Wildlife Biology* (New York: John Wiley & Sons, 1964), pp. 153-56. See also Colin W. Clark, *Mathematical Bioeconomics*, Chapter 7.

4. This latter phenomenon is thought to have been in part responsible for the extinction of the passenger pigeon.

5. The analysis is modified but not fundamentally altered where the tax occurs as a fee levied on the harvest and retained by the system rather than as a restriction that raises costs without generating corresponding revenues.

6. See Harvey Leibenstein, "Bandwagon, Snob, and Veblen Effects in the Theory of Consumers' Demand," *Quarterly Journal of Economics* 64 (May 1950), 183-207. A contemporary example of this

phenomenon was brought to light with the discovery of hunting expeditions to capture the big horn sheep of the Southern California desert. At least thirty-five sportsmen had paid up to $3,500 each for the privilege of being included on "secret safaris" to hunt the species which has been fully protected in California since 1872 [*Audubon* 73 (March 1971), 128]. At present, the mere identification of endangered species is problematical. Although formal listing under the Endangered Species Act offers needed protection for some species, especially on federal lands, it may, for other species, create a fatal notoriety.

7. *Forest and Stream* reported that the increased density of hunters was discouraging the "prudent man" from engaging in deer hunting. In Wisconsin, the deaths of twelve hunters resulted from accidental shooting in 1898 [51 (December 17, 1898), 481].

8. This model ignores the problem of law enforcement. Following a change in the law, the supply schedule would actually shift up by some amount that reflects the expected costs of being caught in violation. The schedule shifts, then, not only with new legislation but with changing levels of enforcement of existing legislation. This is extremely important in the historical case because the cost of legislating restrictions was often close to zero while the cost of securing compliance was large. Although the remainder of this presentation does not explicitly consider the problem of enforcement and its associated costs, these matters are central to the institutional materials treated in the text. Enforcement continues among the weakest elements in wildlife regulation. See Clark S. Bavin, "Wildlife Law Enforcement," in Council on Environmental Quality, *Wildlife and America*, pp. 350-64.

9. Morris D. Johnson, *Feathers from the Prairie: A Short History of Upland Game Birds* (Bismarck: The North Dakota Game and Fish Department, 1964), pp. 62-66. On the controversy over hounding in New York, See Chapter 6, text at notes 49-76.

10. The literature on the growth and regulation of animal populations is voluminous. Useful sources in the preparation of this section include Slobodkin, *Growth and Regulation of Animal Populations*; Robert H. MacArthur and Joseph H. Connell, *The Biology of Populations* (New York: John Wiley & Sons, 1966); McLaren, *Natural Regulation of Animal Numbers*; Dasmann, *Wildlife Biology*; Allen, *Our Wildlife Legacy*; and Edward O. Wilson and William H. Bossert, *A Primer of Population Biology* (Sunderland, MA: Sinauer Associates, 1971).

11. See MacArthur and Connell, *The Biology of Populations*, pp. 135-38.

12. Large population fluctuations seem to be a function of change in environment or of relatively simple population interrelationships. Populations in stable environments are generally able to regulate their numbers within fairly narrow bounds. There have been a number of hypotheses put forth in partial explanation of this ability. They include territoriality, dominance hierarchies, social stress mechanisms, genetic polymorphism, "quality" deterioration accompanying rapid growth, and feedback mechanisms in which the behavior of the group generates density dependent responses in individuals. Discussion and references may be found in Tober, "Allocation of Wildlife Resources," Chapter 2, note 14, pp. 95-97.

13. N'N" need not only be composed of nonsurviving young but also of those born in other periods which die during the year.

14. See Allen, *Our Wildlife Legacy*, passim. The concept of the "huntable surplus" is a major theme of this book.

15. The substitutability of hunting for natural decrements in deer populations is suggested in Raymond Dasmann, *Environmental Conservation*, 3d ed. (New York: John Wiley & Sons, 1972), p. 283. A substantial hunt is followed by a small winter mortality and results in the same breeding population as a gradually declining fall population without a hunt and with a significant winter mortality.

16. Frederick H. Wagner, "Ecosystem Concepts in Fish and Game Management," in George M. Van Dyne, ed., *The Ecosystem Concept in Natural Resource Management* (New York: Academic Press, 1969), pp. 272-73.

BIBLIOGRAPHY

PRIMARY SOURCES

*Government Documents**

California. *Biennial Report of the State Board of Fish Commissioners of the State of California for the Years 1891/92.* Sacramento, 1892.
Colorado. *Report of the State Game and Fish Commissioner.* Denver, 1905/06, 1910/12, 1913/14.
Connecticut. *Records of the Colony and Plantation of New Haven from 1638 to 1649.* Edited by Charles J. Hoadly. Hartford: Case, Tiffany & Co., 1857.
————. *Records of the Colony or Jurisdiction of New Haven, from May, 1653, to the Union.* Edited by Charles J. Hoadly. Hartford: Case, Lockwood & Co., 1858.
————. *The Public Records of the Colony of Connecticut.* 15 vols. Vols. 1-3 edited by J. Hammond Trumbull; Vols. 4-15 edited by Charles J. Hoadly. Hartford, 1850-90.
————. *Records of the State of Connecticut.* 11 vols. Vols. 1-3 edited by Charles J. Hoadly; Vols. 4-8 edited by Leonard Woods Larabee; Vols. 9-10 edited by Albert E. Van Dusen; Vol. 11 edited by Christopher Collier. Hartford, 1894-1967.

*All U.S Government documents cited here are published at Washington, D.C.: U.S. Government Printing Office.

———. *Report of the Commissioners Concerning the Protection of Fish in the Connecticut River, & c.* Hartford, 1867.

———. *Report of the Commissioners of Fisheries.* Hartford, 1868-1894.

———. *Report of the State Commissioners of Fisheries and Game.* Hartford, 1895/96-1899/1900.

Idaho. *Report of the Fish and Game Warden.* [Caldwell?], 1907/08.

Indiana. *Report of the Commissioners of Fisheries and Game.* Indianapolis, 1899/1900-1915/16.

Maine. *Report of the Commissioners of Fisheries.* Augusta, 1867-79.

———. *Report of the Commissioners of Fisheries and Game.* Augusta, 1880-84, 1885/86-1893/94.

———. *Report of the Commissioners of Inland Fisheries and Game.* Augusta, 1895/96-1899/1900.

Massachusetts. *Acts and Resolves, Public and Private, of the Province of the Massachusetts Bay.* 21 vols. Boston: Wright & Potter, Printers to the State, 1869-1922.

———. Commissioners on the Zoological and Botanical Survey. *A Report on the Quadrupeds of Massachusetts,* by Ebenezer Emmons. Cambridge, Mass: Folsom, Wells, and Thurston, 1840.

———. *Report of the Commissioners Appointed under Resolve of 1856, Chap. 58, concerning the Artificial Propagation of Fish.* Boston, 1857.

———. *Report of the Commissioners of Fisheries.* Boston, 1867-85.

———. *Report of the Commissioners on Inland Fisheries and Game.* Boston: 1886-1900.

Michigan. *Report of the State Board of Fish Commissioners.* Lansing, 1873/74-1899/1900.

———. *Report of the State Game and Fish Warden.* Lansing, 1887/88-1899/1900.

Minnesota. *Report of the State Fish Commissioners.* St. Paul, 1874-78, 1879/80-1886/88.

New Hampshire. *Laws of New Hampshire including Public and Private Acts, and Resolves and the Royal Commissions and Instructions.* 7 vols. Vols. 1-2 edited by Albert Stillman Batchellor; Vols. 3-7 edited by H. H. Metcalf. Manchester: The John B. Clarke Co., 1904-18.

———. *Report of the Select Committee on Fisheries.* Concord, 1865.

———. *Report of the Commissioners on Fisheries.* Concord, 1866-82.

———. *Report of the Fish and Game Commissioners.* Concord, 1883-1900.

New Jersey. *Annual Report of the Board of Fish and Game Commissioners.* Trenton, 1897-98.

New York. *The Colonial Laws of New York from the Year 1664 to the Revolution.* 5 vols. Albany: James B. Lyon, State Printer, 1894-96.

———. *Report of the Commissioners of Fisheries.* Albany, 1869-95.

———. *Report of the Commissioners of Fisheries, Game and Forests.* Albany, 1895-1900.

Ohio. *Report of the Ohio State Fish Commissioner.* Columbus, 1875-87.

———. *Report of the Ohio State Fish and Game Commission.* Columbus: 1888/89, 1899/1900.

U.S. Department of Agriculture. *Report of the Commissioner of Agriculture for the Year 1863.* "Dogs and Dog Laws," by J. R. Dodge, pp. 450-63. 1863.

———. *Report of the Commissioner of Agriculture for the Year 1863.* "Agricultural Ornithology: Insectivorous Birds of Chester County, Pennsylvania," by E. Michener, pp. 287-307. 1863.

———. *Report of the Commissioner of Agriculture for the Year 1863.* "Mammology and Ornithology of New England, with Reference to Agricultural Economy," by E. A. Samuels, pp. 269-86. 1863.

———. *Report of the Commissioner of Agriculture for the Year 1864.* "Birds and Bird Laws," by J. R. Dodge, pp. 431-46. 1865.

———. *Report of the Commissioner of Agriculture for the Year 1864.* "The 'Game Birds' of the United States," by D. G. Elliot, pp. 356-85. 1865.

———. *Report of the Commissioner of Agriculture for the Year 1864.* "Oology of Some of the Land Birds of New England, as a Means of Identifying Injurious and Beneficial Species," by E. A. Samuels, pp. 386-430. 1865.

———. *Report of the Commissioner of Agriculture for the Year 1867.* "The Value of Birds on the Farm" by E. A. Samuels, pp. 201-8. 1868.

———. *Report of the Commissioner of Agriculture for the Year 1887.* "Report of the Ornithologist and Mammologist," pp. 402-22. 1888.

———. *Yearbook: 1895.* "Four Common Birds of the Farm and Garden," by Sylvester D. Judd, pp. 405-18. 1896.

———. *Yearbook: 1896.* "Extermination of Noxious Animals by Bounties," by T. S. Palmer, pp 55-68. 1897.

———. *Yearbook: 1898.* "The Dangers of Introducing Noxious Animals and Birds," pp. 87-110. 1899.

———. *Yearbook: 1899.* "Review of Economic Ornithology in the United States," by T. S. Palmer, pp. 259-92. 1900.

———. *Yearbook: 1904.* "Some Benefits the Farmer May Derive from Game Protection," by T. S. Palmer, pp. 509-20. 1905.

———. *Yearbook: 1905.* "Federal Game Protection: A Five Years' Retrospect," by T. S. Palmer, pp. 541-62. 1906.

———. *Yearbook: 1907.* "Bounty Laws in Force in the United States July 1, 1907," by D. E. Lantz, pp. 560-65. 1908.

———. *Yearbook: 1910.* "The Game Market Today," by Henry Oldys, pp. 243-54. 1911.

———. Division of Economic Ornithology and Mammology. *The English Sparrow (Passer domesticus) in North America, Especially in its Relations to Agriculture,* by Walter B. Barrows. Bulletin No. 1. 1889.

———. The Division of Economic Ornithology and Mammology. *The Hawks and Owls of the United States in Their Relation to Agriculture,* by A. K. Fisher. Bulletin No. 3. 1893.

———. Division of Biological Survey. *Digest of Game Laws for 1901,* by T. S. Palmer and H. W. Olds. Bulletin No. 16. 1901.

———. Division of Biological Survey. *Hunting Licenses: Their History, Objects, and Limitations,* by T. S. Palmer. Bulletin No. 19. 1904.

———. Division of Biological Survey. *Coyotes in Their Economic Relations,* by David E. Lantz. Bulletin No. 20. 1905.

———Division of Biological Survey. *The Grouse and Wild Turkeys of the United States, and Their Economic Value,* by Sylvester D. Judd. Bulletin No. 24. 1905.

———. Division of Biological Survey. *The North American Eagles and Their Economic Relations,* by Harry C. Oberholser. Bulletin No. 27. 1906.

———. Division of Biological Survey. *Chronology and Index of the More Important Events in American Game Protection, 1776-1911,* by T. S. Palmer. Bulletin No. 41. 1912.

———Division of Biological Survey. *Index to Papers Relating to the Food of Birds by Members of the Biological Survey in Publications of the United States Department of Agriculture, 1885-1911,* by W. L. McAtee. Bulletin No. 43. 1913.

———. Division of Biological Survey. *Bird Day in Schools,* by T. S. Palmer. Circular No. 17. 1896.

———. Bureau of Biological Survey. *Statistics of Hunting Licenses,* by T. S. Palmer. Circular No. 54. 1906.

———. Bureau of Biological Survey. *Hawks and Owls from the Standpoint of the Farmer,* by A. K. Fisher. Circular No. 61. 1907.

———. Bureau of Biological Survey. *Private Game Preserves and Their Future in the United States,* by T. S. Palmer, Circular No. 72. 1910.

———. Bureau of Biological Survey. *Progress in Game Protection in 1910*, by T. S. Palmer and Henry Oldys. Circular No. 80. 1910.

———. *Game as a National Resource*, by T. S. Palmer. Bulletin No. 1049. 1922.

U.S. Patent Office. *Report of the Commissioner of Patents for 1856: Agriculture.* "Birds Injurious to Agriculture," by Ezekial Holmes, pp. 110-60. 1856.

———. *Report of the Commissioner of Patents for 1856: Agriculture.* "Quadrupeds of Illinois Injurious and Beneficial to Agriculture," by Robert Kennicott, pp.52-110. 1856. Parts two and three of this study appear in the *Reports* for the years 1857 and 1858, pp. 72-107 and pp. 241-56, respectively.

Vermont. *Report of the Fish Commissioners*. Montpelier, 1867-92.

———. *Report of the Commissioners of Fish and Game*. Montpelier, 1894-1900.

Wisconsin. *Report of the Commissioners of Fisheries*. Madison, 1874-92.

———. *Report of the Commissioners of Fisheries and State Fish and Game. Warden*. Madison: 1893/94-1899/1900.

Periodicals

American Sportsman. 10 vols. West Meriden, Connecticut, 1871-74, and New York 1874-77. Published as *The Rod and Gun and American Sportsman* from 1875 to 1877. Merged into *Forest and Stream*, 1877.

Forest and Stream. 55 vols. New York, 1873-1900.

Recreation: A Monthly Exponent of the Higher Literature of Manly Sport. 3 vols. Chicago, 1888-89. Vols. 1-2 published as *Wildwood's Magazine*.

Recreation. 12 vols. New York, 1894-1900.

Western Field and Stream. 5 vols. St. Paul and New York, 1896-1900. Published as *Field and Stream* from 1898. Absorbed *Forest and Stream* in 1930.

Books and Articles

A Gentleman. *The Sportsman's Companion; or, an Essay on Shooting*. New York: Robertsons, Mills and Hicks, 1783.

A Gentleman of Philadelphia County. *The American Shooter's Manual* (1827). New York: Ernest R. Gee, 1928.

Alexander, Thomas. *Game Birds of the United States: Their Habits and Haunts*. The Seaside Library. Vol. 28, No. 571. New York: George Munro, 1879.

Allen, J. A. "The Former Range of Some N. E. Carnivorous Animals." *American Naturalist* 10 (December 1876), 708-15.

———. "Decrease of Birds in Massachusetts." *Bulletin of the Nuttall Ornithological Club* 1 (September 1876), 53-60.

———. "The Present Wholesale Destruction of Bird-Life in the United States." *Science* 7 (February 26, 1886), 191-95.

American Game Protective and Propagation Association. *The Game of a Continent, Ours to Protect* (1913?).

An Attorney. *A view of the Principal Parts of the Most Important Statutes Relating to Game: With Explanatory Cases and Observations.* London: J. Ellis, 1801.

Anonymous. *A Description of Georgia, by a Gentleman who has Resided there upwards of seven years, and was One of the First Settlers.* (1741), vol. 2. In *Tracts and Other Papers, Relating Principally to the Origin, Settlement, and Progress of the Colonies in North America, from the Discovery to the Year 1776.* 4 vols. Collected by Peter Force. Washington: Printed by Peter Force, 1838.

Anonymous. "Nuttall Ornithological Club." *Bulletin of the Nuttall Ornithological Club* 1 (July 1876), 29-32.

Babcock, Charles A. *Bird Day: How to Prepare for It.* New York: Silver, Burdett and Company, 1901.

Bachelder, John B. *Popular Resorts and How to Reach Them.* 2d ed. Boston: John B. Bachelder, 1874.

Barker, Fred C., and J. S. Danforth. *Hunting and Trapping on the Upper Magalloway River and Parmachenee Lake.* Boston: D. Lothrop, 1882.

Beltrami, G. C. *A Pilgrimage in Europe and America, leading to the Discovery of the Sources of the Mississippi and Bloody River.* 2 vols. London: Hunt and Clarke, 1828.

Blackstone, William. *Commentaries on the Laws of England, together with Notes Adapting the Work to the American Student.* By John L. Wendell. 3 vols. New York: Harper and Brothers, 1847.

Bogardus, Adam H. *Field, Cover, and Trap Shooting.* 2d ed. New York: Published by the Author, 1878.

Boone and Crockett Club. *Officers, Constitution, and List of Members for the Year 1907.*

———. Report of the Game Preservation Committee, 1915.

Bradley, A. G. "Game Preserving in the United States." *Macmillian's Magazine* 58 (1888), 364-70.

Brownell, L. W. *Photography for the Sportsman Naturalist.* New York: Macmillan Co. 1904.

Browning, Meshach. *Forty-four Years of the Life of a Hunter*. Philadelphia: J. B. Lippincott & Co., 1864.

Bumstead, John. *On the Wing: A Book for Sportsmen*. Boston: Fields, Osgood, and Co., 1869.

Cartwright, David W. *Natural History of Western Wild Animals, and Guide for Hunters, Trappers, and Sportsmen*. Toledo: Blade Printing and Paper Co., 1875.

Catlin, George. *Letters and Notes on the Manners, Customs, and Conditions of the North American Indians*. 2 vols. Minneapolis: Ross and Haynes, 1965.

Caton, John Dean. *The Antelope and Deer of America: A Comprehensive Scientific Treatise upon the Natural History, including the Characteristics, Habits, Affinities, and Capacity for Domestication of the Antilocapra and Cervidae of North America*. 2d ed. New York: Forest and Stream Publishing Company, 1881.

Chapman, Frank M. *Bird Studies with a Camera*. New York: D. Appleton & Co., 1900.

Chase, Henry. *Game Protection and Propagation in America*. Philadelphia: J. B. Lippincott Company, 1913.

Cobbett, William. *A Year's Residence in the United States of America*. 2d ed. London: Sherwood, Neely, & Jones, 1819.

Cody, Wiliam F. *True Tales of the Plains*. New York: Cupples and Leon Company, 1908.

Colles, James. *Journal of a Hunting Excursion to Louis Lake, 1851*. Blue Mountain Lake, New York: Adirondack Museum, 1961.

Crèvecoeur, J. Hector St. John. *Letters from an American Farmer*. New York: Fox, Duffield & Co., 1904.

Davis, Gherardi. *The Southside Sportsmen's Club of Long Island*. Privately printed, 1909.

De Voe, Thomas F. *The Market Assistant, Containing a Brief Description of Every Article of Human Food Sold in the Public Markets of the Cities of New York, Boston, Philadelphia, and Brooklyn*. New York: Orange Judd Publishing Company, 1866.

Dodge, Richard Irving. *The Plains of the Great West and their Inhabitants, being a Description of the Plains, Game, and Indians, & c. of the Great North American Desert*. New York: G. P. Putnam's Sons, 1877.

Doughty, T. "Characteristics of a True Sportsman." Reprinted in *Classics of the American Shooting Field: A Mixed Bag for the Kindly Sportsman, 1783-1926*, edited by John C. Phillips and Lewis Webb Hill, M.D. Boston: Houghton Mifflin Co., 1930.

Dugmore, A. R. *Nature and the Camera*. New York: Doubleday, Page & Co., 1903.

Dutcher, William. "Destruction of Bird-Life in the Vicinity of New York." *Science* 7 (February 26, 1886), 197-99.

Ellsworth, Lincoln. *The Last Wild Buffalo Hunt*. New York: Privately printed, 1916.

Fur, Fin and Feather. A Compilation of the Game Laws of the Different States and Provinces of the United States and Canada; to which is added a List of Hunting and Fishing Localities, and other Useful Information for Gunners and Anglers. Revised and corrected for 1871-72. New York: M. B. Brown & Co., 1871.

Godman, John D. *American Natural History*. 2 vols. 3d ed. Philadelphia: Hogan and Thompson, 1836.

Goodhue, Stoddard. "Game Laws and Game of America." *Harper's Weekly* 40 (February 1, 1896), 118; 40 (February 8, 1896), 141-42.

Grant, Madison. "Reform in the New York Game Laws." *Harper's Weekly* 40 (October 3, 1896), 978.

Greeley, William B. "Reform in Game Administration." *Transactions of the Fifteenth National Game Conference*, pp. 89-97. American Game Protective Association, 1928.

Greener, W. W. *The Gun and Its Development*. London: Cassell, Petter, Galpin & Co., 1881.

Grinnell, George Bird. *American Duck Shooting*. New York: Forest and Stream Publishing Company, 1901.

———. "Brief History of the Boone and Crockett Club." In *Hunting at High Altitudes*, edited by George B. Grinnell. New York: Harper & Brothers, 1913.

———. "American Game Protection: A Sketch." In *Hunting and Conservation. The Book of the Boone and Crockett Club*, edited by George Bird Grinnell and Charles Sheldon. New Haven: Yale University Press, 1925.

Hadley, Alden H. "The Legal Status of Hawks and Owls: A Statistical Study." *Transactions of the Fifteenth National Game Conference*, pp. 41-48. American Game Protective Association, 1928.

Hallock, Charles, *The Fishing Tourist: Angler's Guide and Reference Book*. New York: Harper & Brothers, 1873.

———. *Hallock's American Club List and Sportsman's Glossary*. New York: Forest and Stream Publishing Company, 1878.

———. *The Sportsman's Gazeteer and General Guide*. 5th ed. New York: Forest and Stream Publishing Company, 1879.

———. *An Angler's Reminiscences: A Record of Sport, Travel and*

Adventure, with Autobiography of the Author. Cincinnati: Sportsmen's Review Publishing Co., 1913.

Harris, William C., comp. *The Sportsman's Guide to the Hunting and Shooting Grounds of the United States and Canada.* New York: The Anglers' Publishing Company, Chas. T. Dillingham, 1888.

Hays, W. J. "Notes on the Range of Some of the Animals in America at the Time of the Arrival of the White Men." *American Naturalist* 5 (September 1871), 387-92.

Herbert, Henry William. *Field Sports in the United States, and the British Provinces of America.* 2 vols. London: Richard Bentley, 1848.

Hoar, George F. *Autobiography of Seventy Years.* 2 vols. New York: Charles Scribner's Sons, 1903.

Holberton, Wakeman. "The Supply of Game: Influence of Clubs and Private Game Preserves." *Harper's Weekly* 37 (April 8, 1893), 339.

Hooper, Johnson J. *Dog and Gun.* New York: Orange Judd Publishing Company, 1869.

Hornaday, William T. *The Extermination of the American Bison: with a Sketch of Its Discovery and Life History.* The Annual Report of the U.S. National Museum. Washington, D.C., 1887.

———. *The Destruction of Our Birds and Mammals.* Extracted from the Second Annual Report of the New York Zoological Society. New York, 1898.

———. *Our Vanishing Wildlife: Its Extermination and Preservation.* New York: The New York Zoological Society, 1913.

Hunter, Alexander. "The Club-Houses of Currituck Sound." *Harper's Weekly* 36 (March 12, 1892), 353-55.

Ingham, John H. *The Law of Animals: A Treatise on Property in Animals Wild and Domestic and the Rights and Responsibilities Arising Therefrom.* Philadelphia: T. and J. W. Johnson & Co. 1900.

Jefferson, Thomas. *Notes on the State of Virginia.* Edited with an Introduction and Notes by William Peden. Chapel Hill: University of North Carolina Press, 1955.

Johnson, T. B. *The Gamekeeper's Directory, and Complete Vermin Destroyer.* London: Sherwood, Gilbert, and Piper, 1832.

Josselyn, John. *New England's Rarities Discovered* (1672). *Archaeologia Americana. Transactions and Collections of the American Antiquarian Society,* 12 vols (Worcester: Printed for the Society, 1820-1911), IV.

Judd, David W., ed., *Life and Writings of Frank Forester,* 2 vols. New York: Orange Judd Publishing Co., 1882.

Kennebec Association for the Protection of Fish and Game. *A Catalogue of the Officers and Members, Act of Incorporation, Constitution, By-Laws and Obligations, The President's Address, and Fish and Game Laws.* Augusta, Me.: George E. Nason's Job Printing Office, 1878.

Lawson, John. *The History of Carolina* (1714). Raleigh: Strother and Marcom, 1860.

Leopold, Aldo. *Report on A Game Survey of the North Central States.* Madison, Wis.: Sporting Arms and Ammunition Manufacturers' Institute, Committee on Restoration and Protection of Game, 1931.

————. *Game Management.* New York: Charles Scribner's Sons, 1933.

Lewis, E. J. *Hints to Sportsmen, Containing Notes on Shooting; The Habits of the Game Birds and Wild Fowl of America; the Dog, the Gun, the Field and the Kitchen.* Philadelphia: Lea and Blanchard, 1851.

Marsh, George Perkins. *Man and Nature* (1864). Edited by David Lowenthal. Cambridge, Mass.: Belknap Press of Harvard University Press, 1965.

Mather, Fred. *My Angling Friends.* New York: Forest and Stream Publishing Co., 1901.

Merritt, H. Clay. *The Shadow of a Gun.* Chicago: F. T. Peterson Company, 1904.

Mershon, W. B., ed. *The Passenger Pigeon.* New York: The Outing Publishing Company, 1907.

Morton, Thomas. *New English Canaan; or New Canaan, Containing an Abstract of New England.* (1632), vol. 2. In *Tracts and Other Papers.* 4 vols. Collected by Peter Force. Washington, D.C.: Printed by Peter Force, 1838.

Moulton, Joseph W. *An Address Delivered at St. Paul's Church, Buffalo, on the Anniversary Celebration of the Niagara and Erie Society for Promoting Agriculture and Domestic Manufactures.* Buffalo: Printed by D. M. Day and H. A. Salisbury, 1821.

Newhouse, S. *The Trapper's Guide.* 2d. ed. Wallingford, Conn.: Oneida Community, 1867.

New York Association for the Protection of Game. *Constitution and List of Officers and Members of the New York Association for the Protection of Game,* 1891.

Nuttall, Thomas. *A Manual of Ornithology: The Land Birds.* Boston: Hilliard, Gray & Co., 1840.

Pearson, Thomas Gilbert. *Adventures in Bird Protection: An Autobiography.* New York: D. Appleton-Century Company, 1937.

Poland, Henry. *Fur-Bearing Animals in Nature and Commerce*. London: Gurney and Jackson, 1892.

Robinson, Solon, ed. *Facts for Farmers*. 2 vols. New York: A. J. Johnson, 1866.

Roosevelt, Robert B. *The Game Birds of the Coasts and Lakes of the Northern States of America*. New York: Carleton, 1866.

Roosevelt, Theodore. *The Wilderness Hunter*. New York: G. P. Putnam's Sons, 1893.

————. "The Boone and Crockett Club." *Harper's Weekly* 37 (March 18, 1893), 267.

Sanger, William Cary. "The Adirondack Deer Law." In *Hunting and Conservation. The Book of the Boone and Crockett Club*, edited by George Bird Grinnell and Charles Sheldon. New Haven: Yale University Press, 1925.

Schreiner, William H. *Schreiner's Sporting Manual*. Philadelphia: S. D. Wyeth, 1841.

Scott, W.E.D. "The Present Condition of Some of the Bird Rookeries of the Gulf Coast of Florida." *The Auk* 4 (1887), 135-44, 213-27, 273-84.

Seccomb, Joseph. "Business and Diversion Inoffensive to God, and Necessary for the Comfort and Support of Human Society. A discourse utter'd in part at Ammauskeeg Falls, [New Hampshire] in the fishing season, 1739." Boston: Printed for S. Kneeland and T. Green in Queen-Street, 1743; Reprinted. Manchester, N.H.: 1892.

Sennett, Geo. B. "Destruction of the Eggs of Birds for Food." *Science* 7 (February 26, 1886), 199-201.

Shields, G. O., ed. *The Big Game of North America*. Chicago: Rand, McNally & Co., 1890.

Shiras, George, III. *Hunting Wild Life with Camera and Flashlight*. 2d ed. 2 vols. Washington, D.C.: National Geographic Society, 1936.

Siders, James Buchanan. *The Nimrods or, How to Hunt and Shoot*. Dayton, Ohio: (1905?).

Smith, George Putnam. *The Law of Field Sports*. New York: Orange Judd Publishing Co., 1886.

Soulé, H. H. *Hints and Points for Sportsmen*. New York: Forest and Stream Publishing Co., 1889.

Stoddard, S. R. *The Adirondacks Illustrated*. 28th ed. Glens Falls, N.Y.: Published by the Author, 1898.

South Side Sportsmen's Club of Long Island. *Constitution and By-Laws*. New York: Evening Post Steam Presses. 1874.

Sweetser, Charles H. *Tourists' and Invalids' Guide to the Northwest.* New York: Evening Mail, 1868.

Thomas, William S. *Hunting Big Game with Gun and Kodak.* New York: G. P. Putnam's Sons, 1906.

Titcomb, John W. "Fish and Game in Vermont." In *With Rod and Gun in New England and the Maritime Provinces,* edited by Edward A. Samuels. Boston: Samuels and Kimball, 1897.

Tome, Phillip. *Pioneer Life; or Thirty Years a Hunter* (1854). Harrisburg, Pa.: The Aurand Press, 1928.

Van Dyke, Henry. *Fisherman's Luck.* New York: Charles Scribner's Sons, 1899.

Warden, D. B. *A Statistical, Political, and Historical Account of the United States of America.* 3 vols. Edinburgh: Archibald Constable & Co., 1819.

Webb, W. E. *Buffalo Land: An Authentic Account of the Discoveries, Adventures, and Mishaps of a Scientific and Sporting Party in the Wild West.* Cincinnati: E. Hannaford & Co., 1872.

Welford, Richard Griffiths. *The Influences of the Game Laws: Being Classified Extracts from the Evidence Taken before a Select Committee of the House of Commons on the Game Laws, with an Appendix and An Address to the Tenant Farmers of Great Britain by John Bright, Esq. M.P.* London: Groombridge and Sons, 1846.

Werich, J. Lorenzo. *Pioneer Hunters of the Kankakee.* Printed by the Author, 1920.

West Jersey Game Protective Society. *Act of Incorporation and Constitution and By Laws of the West Jersey Game Protective Society and New Jersey Game Laws,* 1876.

Wildwood, Will [Fred E. Pond], comp. *The Sportsman's Directory.* Milwaukee: Fred E. Pond, 1891.

Wilson, Alexander. *American Ornithology.* Edited by T. M. Brewer. Boston: Otis, Broaders, and Company, 1840.

SECONDARY SOURCES

Allen, Durward L. *Our Wildlife Legacy.* Rev. ed. New York: Funk & Wagnals, 1962.

Allen, Glover M. *Extinct and Vanishing Mammals of the Western Hemisphere.* Special Publication #11. New York: American Committee for International Wildlife Protection, 1942.

Amory, Cleveland. *Man Kind? Our Incredible War on Wildlife*. New York: Harper and Row, 1974.

Barnett, Harold J., and Chandler Morse. *Scarcity and Growth: The Economics of Natural Resource Availability*. Baltimore: Published for Resources for the Future by Johns Hopkins University Press, 1963.

Bavin, Clark R. "Wildlife Law Enforcement." In Council on Environmental Quality, *Wildlife and America*, pp. 350-64.

Bersing, Otis S. *A Century of Wisconsin Deer*. 2d ed. Madison, Wis.: Game Management Division of the Wisconsin Conservation Department, 1966.

Billington, Ray Allen. *The Far Western Frontier, 1830-1860*. New York: Harper & Row, 1956.

Bolle, Arnold W., and Richard D. Taber. "Economic Aspects of Wildlife Abundance on Private Lands." *Transaction of the North American Wildlife Conference*. pp. 255-67. Washington, D.C.: American Wildlife Institute, 1962.

Boyd, William S. "Federal Protection of Endangered Wildlife Species." *Stanford Law Review* 22 (June 1970), 1289-1309.

Branch, Douglas, E. *The Hunting of the Buffalo*. New York: D. Appleton and Co., 1929.

Brown, Dee. *Bury My Heart at Wounded Knee: An Indian History of The American West*. New York: Bantam Books, 1971.

Bruce, Philip Alexander. *Economic History of Virginia in the Seventeenth Century*. 2 vols. New York: Macmillan & Co., 1896: New York: Johnson Reprint Corp., 1966.

Buchheister, Carl W., and Frank Graham, Jr. "From the Swamps and Back—A Concise and Candid History of the Audubon Movement." *Audubon* 75 (January 1973) 4-45.

Bushman, Richard L. *From Puritan to Yankee*. Cambridge, Mass: Harvard University Press, 1967.

Cameron, Jenks. *The Bureau of Biological Survey: Its History, Activities and Organization*. Service Monograph of the United States Government, No. 54. Baltimore: Institute for Government Research of the Brookings Institution by Johns Hopkins Press, 1929.

Carstensen, Vernon, ed. *The Public Lands: Studies in the History of the Public Domain*. Madison: University of Wisconsin Press, 1968.

Cart, Theodore Whaley. "The Lacey Act: America's First Nationwide Wildlife Statute." *Forest History* 17 (October 1973), 4-13.

————. "The Struggle for Wildlife Protection in the United States,

1870-1900: Attitudes and Events Leading to the Lacey Act."
Ph.D. diss., University of North Carolina, 1971.

Chancellor, John. *Audubon: A Biography*. New York: The Viking Press,
1978.

Charne, Irvin B. "Fish and Game—Power of the State to Regulate
Taking Of." *Wisconsin Law Review* (January 1949), 181-84.

Chittenden, Hiram Martin. *The American Fur Trade of the Far West*. 2
vols. New York: The Press of the Pioneers, 1935.

Clark, Colin W. *Mathematical Bioeconomics: The Optimal Management of
Renewable Resources*. New York: John Wiley & Sons, 1976.

———. "Profit Maximization and the Extinction of Animal Populations." *Journal of Political Economy* 81 (1973), 950-61.

Coggins, George Cameron. "Legal Protection for Marine Mammals;
An Overview of Innovative Resource Conservation Legislation."
Environmental Law 6 (1975), 1-59.

——— and William H. Hensley. "Constitutional Limits on Federal
Power to Protect and Manage Wildlife: Is the Endangered
Species Act Endangered?" *Iowa Law Review* 61 (June 1976),
1099-1152.

——— and Deborah Lyndall Smith. "The Emerging Law of Wildlife: A
Narrative Bibliography." *Environmental Law* 6 (1975), 583-618.

Colorado Department of Natural Resources, Division of Game, Fish
and Parks. *A Look Back: A 75 Year History of the Colorado Game,
Fish and Parks Division*. Denver, 1972.

Connery, Robert H. *Governmental Problems in Wild Life Conservation*.
New York: Columbia University Press, 1935.

Council on Environmental Quality. *The Evolution of National Wildlife
Law*, by Michael J. Bean. Washington, D.C.: G.P.O., 1977.

———. *Wildlife and America: Contributions to an Understanding of American Wildlife and Its Conservation*. Howard P. Brokaw, Project
Director and Editor. Washington, D.C.,: G.P.O., 1978.

Craig, Sandra J. "Wildlife in the National Parks." *Natural Resources
Journal* 12 (October 1972), 627-32.

Crane, Verner W. *The Southern Frontier, 1670-1732*. Philadelphia:
University of Pennsylvania Press, 1929.

Dahlberg, Burton L., and Ralph C. Guettinger. *The White-tailed Deer in
Wisconsin*. Technical Wildlife Bulletin Number 14. Madison:
Game Management Division of the Wisconsin Conservation
Department, 1956.

Dance, S. Peter. *The Art of Natural History: Animal Illustrators and Their
Work*. Woodstock, N.Y.: Overlook Press, 1978.

Dasmann, Raymond. *Wildlife Biology*. New York: John Wiley & Sons, 1964.

———. *Environmental Conservation*. 3d ed. New York: John Wiley & Sons, 1972.

Davis, Lance E., and Douglass C. North. *Institutional Change and American Economic Growth*. Cambridge: At the University Press, 1971.

Day, Gordon M. "The Indian as an Ecological Factor in the Northeastern Forest." *Ecology* 34 (April 1953), 329-46.

Demsetz, Harold. "Toward a Theory of Property Rights." *American Economic Review* 62 (May 1967), 347-59.

Dickens, James R. "The Law and Endangered Species of Wildlife." *Gonzaga Law Review* 9 (Fall 1973), 57-115.

Dickenson, Nathanial, and Lawrence E. Garland. *The White-tailed Deer Resources of Vermont*. Montpelier: The Vermont Fish and Game Department, Agency of Environmental Conservation, 1974.

Doughty, Robin W. *Feather Fashions and Bird Preservation: A Study in Nature Protection*. Berkeley: University of California Press, 1975.

Dulles, Foster Rhea. *America Learns to Play: A History of Popular Recreation, 1607-1940*. Gloucester: Peter Smith, 1959. Reprint of 1940 edition.

Durham, H. M., and R. B. Herrington. "State Ownership of Fish and Game." *Georgetown Law Review* 38 (1950), 652-58.

Ehrenfeld, David W. *Biological Conservation*. New York: Holt, Rinehart and Winston, 1970.

Ekirch, Arthur. *The Idea of Progress in America, 1815-1860*. New York: Columbia University Press, 1944.

Ellison, Joseph. "The Mineral Land Question in California, 1848-1866." *Southwestern Historical Quarterly* 30 (1926), 34-55. Reprinted in Carstensen, ed., *The Public Lands*.

Englebert, Ernest A. "American Policy for Natural Resources: A Historical Survey to 1862." Ph.D. diss. Harvard University, 1950.

Farb, Peter. *The Land and Wildlife of North America*. A Volume in the Life Nature Library. New York: Time Incorporated, 1966.

———. *Man's Rise to Civilization as Shown by the Indians of North America from Primeval Times to the Coming of the Industrial State*. New York: E. P. Dutton & Co., Inc., 1968.

Firey, Walter. *Man, Mind, and Land: A Theory of Resource Use*. New York: The Free Press, 1960.

Flader, Susan L. *Thinking Like a Mountain*. Columbia: University of Missouri Press, 1974.

―――. "Scientific Resource Management: An Historical Perspective." *Transactions: Forty-first North American Wildlife and Natural Resources Conference*. Wildlife Management Institute, 1976.

Foss, Phillip O., ed. *Conservation in the United States: A Documentary History—Recreation*. New York: Chelsea House Publishers, 1971.

Friedman, Lawrence M. *A History of American Law*. New York: Simon and Schuster, 1973.

Frome, Michael. "Panthers Wanted Alive, Back East Where They Belong." *Smithsonian* 10 (June 1979), 83-87.

Gabriel, Ralph Henry. *The Course of American Democratic Thought: An Institutional History Since 1815*. New York: The Ronald Press, 1940.

Gard, Wayne. *The Great Buffalo Hunt*. New York: Alfred A. Knopf, 1959.

Garretson, Martin S. *The American Bison*. New York: New York Zoological Society, 1938.

Gilmore, Grant. *The Ages of American Law*. New Haven: Yale University Press, 1977.

Goodspeed, Charles Eliot. *Angling in America, Its Early History and Literature*. Boston: Houghton Mifflin Company, 1939.

Gordon, H. Scott. "The Economic Theory of A Common Property Resource: The Fishery." *Journal of Political Economy* 62 (April 1954), 124-42.

Gottschalk, John S. "The State-Federal Partnership in Wildlife Conservation." In Council on Environmental Quality, *Wildlife and America*, pp. 290-301.

Graham, Frank, Jr. *The Adirondack Park: A Political History*. New York: Alfred A. Knopf, 1978.

Greenwalt, Lynn A. "The National Wildlife Refuge System." In Council on Environmental Quality, *Wildlife and America*, pp. 399-412.

Greenway, James C., Jr. *Extinct and Vanishing Birds of the World*. Special Publication No. 13. New York: American Committee for International Wildlife Protection, 1958.

Hanley, Wayne. *Natural History in America, from Mark Catesby to Rachel Carson*. New York: Quadrangle/New York Times Book Company, 1977.

Hardin, Garrett. "Political Requirements for Preserving Our Common Heritage." In Council on Environmental Quality, *Wildlife and America*. pp. 310-17.

————. "The Tragedy of the Commons." *Science* 162 (1968), 1243-48.

Harding, T. Swann. *Two Blades of Grass: A History of Scientific Development in the U.S. Department of Agriculture.* Norman: University of Oklahoma Press, 1947.

Harris, Neil. *Humbug: The Art of P.T. Barnum.* Boston: Little Brown, 1973.

Hays, Samuel P. *Conservation and the Gospel of Efficiency: The Progressive Conservation Movement, 1890-1920.* Cambridge, Mass.: Harvard University Press, 1959.

Hester, James J. "The Agency of Man in Animal Extinctions." In *Pleistocene Extinctions: The Search for a Cause,* edited by P. S. Martin and H. E. Wright, Jr. New Haven: Yale University Press, 1967.

Hewitt, C. Gordon. *The Conservation of Wildlife of Canada.* New York: Charles Scribner's Sons, 1921.

Horwitz, Morton J. "The Transformation in the Conception of Property in American Law, 1780-1860." *University of Chicago Law Review* 40 (1973). Reprinted in *American Law and the Constitutional Order: Historical Perspective,* edited by Lawrence M. Friedman and Harry N. Scheiber. Cambridge, Mass.: Harvard University Press, 1978.

————. *The Transformation of American Law, 1790-1860.* Cambridge, Mass.: Harvard University Press, 1977.

Huth, Hans. *Nature and the American: Three Centuries of Changing Attitudes.* Berkeley: University of California Press, 1957.

Ise, John. *Our National Park Policy.* Baltimore: Published for Resources for the Future by Johns Hopkins Press, 1961.

Jelinek, Arthur J. "Man's Role in the Extinction of Pleistocene Faunas." In *Pleistocene Extinctions: The Search for a Cause,* edited by P. S. Martin and H. E. Wright, Jr. New Haven: Yale University Press, 1967.

Johnson, Morris D. *Feathers from the Prairie: A Short History of Upland Game Birds.* Bismarck: North Dakota Game and Fish Department, 1964.

Josephy, Alvin M. *The Indian Heritage of America.* New York: Alfred A. Knopf, 1968.

Kane, Lucile. "Federal Protection of Public Timber in the Upper Great Lakes States." *Agricultural History* 23 (1949), 135-39. Reprinted in Carstensen, ed., *The Public Lands,* pp. 439-47.

Kastner, Joseph. *A Species of Eternity.* New York: E. P. Dutton, 1978.

Kimball, David, and Jim Kimball. *The Market Hunter.* Minneapolis: Dillon Press, 1969.

Kimball, Thomas L., and Raymond E. Johnson. "The Richness of American Wildlife." In Council on Environmental Quality, *Wildlife and America*, pp. 3-17.

King, F. Wayne. "The Wildlife Trade." In Council on Environmental Quality, *Wildlife and America*, pp. 253-71.

Kirby, Chester. "The English Game Law System." *American Historical Review* 38 (January 1933), 240-62.

Knight, Rolf. "A Reexamination of Hunting, Trapping, and Territoriality among the Northeastern Algonkian Indians." In *Man, Culture, and Animals*, edited by A. Leeds and A. P. Vayda. Washington, D.C.: American Association for the Advancement of Science, 1965.

Kobey, Eugene F. "Discrimination by States Against Non-Residents' Hunting and Fishing Privileges." *Marquette Law Review* 33 (1950), 92-95.

Kranz, Marvin W. "Pioneering in Conservation: A History of the Conservation Movement in New York State, 1865-1903." Ph.D. diss., Syracuse University, 1961.

Laycock, George. *The Alien Animals*. Garden City, N.Y.: Published for the American Museum of Natural History by the Natural History Press, 1966.

Libecap, Gary D. "Economic Variables and the Development of the Law: The Case of Western Mineral Rights." *Journal of Economic History* 38 (June 1978), 338-62.

Lipton, James. *An Exaltation of Larks, or the Venereal Game*. New York: Grossman, 1968.

Lockridge, Kenneth A. *A New England Town: The First Hundred Years, Dedham, Massachusetts, 1636-1736*. New York: W. W. Norton, 1970.

Lofgren, Charles A. "*Missouri v. Holland* in Historical Perspective." *The Supreme Court Review* (1975), 71-122.

Lund, Thomas A. "British Wildlife Law Before the American Revolution: Lessons From the Past." *Michigan Law Review* 74 (November 1975), 49-74.

———. "Early American Wildlife Law." *New York University Law Review* 51 (November 1976), pp. 703-30.

McArthur, Robert H., and Joseph H. Connell. *Biology of Populations*. New York: John Wiley & Sons, 1966.

McCoy, J. J. *Saving Our Wildlife*. New York: Crowell-Collier Press, 1970.

McHugh, Tom. "Bison Travels, Bison Travails." *Audubon* 74 (November 1972), 22-31.

McLaren, Ian A., ed. *Natural Regulation of Animal Populations.* New York: Atherton Press, 1971.

Manchester, Herbert. *Four Centuries of Sport in America, 1490-1890.* New York: The Derrydale Press, 1931.

Martin, Calvin. *Keepers of the Game: Indian Animal Relationships and the Fur Trade.* Berkeley, University of California Press, 1978.

Martin, Paul S. "Prehistoric Overkill." In *Pleistocene Extinctions: The Search for a Cause,* edited by P. S. Martin and H. E. Wright, Jr. New Haven: Yale University Press, 1967.

Marx, Leo. *The Machine in the Garden: Technology and the Pastoral Ideal in America.* New York: Oxford University Press, 1964.

Mattheissen, Peter. *Wildlife in America.* New York: Viking Press, 1959.

Merriman, R. O. *The Bison and the Fur Trade.* Bulletin of the Department of History and Political and Economic Science in Queen's University, #53. Kingston, Ontario: 1926.

Miller, John C. *The First Frontier: Life in Colonial America.* New York: Delacorte Press, 1966.

Miller, Jon R. and Fredric C. Menz. "Some Economic Considerations for Wildlife Preservation." *Southern Economic Journal* 45 (January 1979), 718-29.

Miller, Richard S., and Daniel B. Botkin. "Endangered Species: Models and Predictions." *American Scientist* 62 (March-April 1974), 172-81.

Morgan, Edmund S. *The Puritan Dilemma: The Story of John Winthrop.* Boston: Little, Brown and Company, 1958.

Nash, Roderick. *Wilderness and the American Mind.* New Haven: Yale University Press, 1967.

———, ed. *The American Environment: Readings in the History of Conservation.* Reading, Mass.: Addison-Wesley Publishing Company, 1968.

Nelson, William E. *Americanization of the Common Law: The Impact of Legal Change on Massachusetts Society, 1760-1830.* Cambridge, Mass: Harvard University Press, 1975.

Nettels, Curtis P. *The Emergence of a National Economy.* New York: Harper & Row, 1962.

Olds, Nicholas V., and Harold W. Glassen. "Do States Still Own Their Game and Fish?" *Michigan State Bar Journal* 30 (1951), 16-23.

Olson, Mancur. *The Logic of Collective Action.* Cambridge, Mass: Harvard University Press, 1965.

Osgood, Ernest Staples. *The Day of the Cattleman.* Chicago: University of Chicago Press, 1957. Reprint of 1929 edition.

Parker, William N. "The Land, Minerals, Water, and Forests." In Lance E. Davis, *et. al.*, *American Economic Growth, an Economists' History of the United States*. New York: Harper and Row, 1972.

Peattie, Donald Culross. *Green Laurels: The Lives and Achievements of the Great Naturalists*. New York: Simon and Schuster, 1936.

Peters, Harold S. "The Past Status and Management of the Mourning Dove." *North American Wildlife Conference* 1961, 371-74.

Peterson, Eugene T. "The History of Wildlife Conservation in Michigan, 1859-1921." Ph.D. diss., University of Michigan, 1952.

Peterson, Frederick M., and Anthony C. Fisher. "The Exploitation of Extractive Resources: A Survey." *Economic Journal* 87 (December 1977), 681-721.

Peterson, Roger Tory. "The Evolution of A Magazine." *Audubon* 75 (January 1973), 46-51.

Phillips, John C. *Migratory Bird Protection in North America: The History of Control by the United States Federal Government and a Sketch of the Treaty with Great Britain*. Special Publication of the American Committee for International Wildlife. Vol. I, No. 4, 1934.

Pinney, Roy. *Vanishing Wildlife*. New York: Dodd, Mead, & Co., 1963.

Poole, Austin Lane. *From Domesday Book to Magna Carta, 1087-1216*, 2d ed. Oxford: The Clarendon Press, 1955.

Potter, David. *People of Plenty; Economic Abundance and The American Character*. Chicago: University of Chicago Press, 1954.

Ray, Arthur J. *Indians in the Fur Trade: Their Role as Trappers, Hunters and Middlemen in the Lands Southwest of Hudson Bay, 1660-1870*. Toronto: University of Toronto Press, 1974.

Regenstein, Lewis. *The Politics of Extinction: The Shocking Story of the World's Endangered Wildlife*. New York: Macmillan Publishing Co., 1975.

Reiger, George. "Hunting and Trapping in the New World." In Council on Environmental Quality, *Wildlife and America*, pp. 42-52.

——, "The Song of the Seal." *Audubon* 77 (September 1975), 6-27.

Reiger, John F. "George Bird Grinnell and the Development of American Conservation." Ph.D. diss., Northwestern University, 1970.

——. *American Sportsmen and the Origins of Conservation*. New York: Winchester Press, 1975.

——, ed. *Passing of the Great West: Selected Papers of George Bird Grinnell*. New York: Winchester Press, 1972.

Riggs, Dan. "Constitutional Law - Wyoming's Guide Law: Nonresident Hunters on Public Lands and Collateral Issues." *Land and Water Law Review* (1974), 169-83.

Robinson, Glen O. *The Forest Service*. Baltimore: Johns Hopkins University Press, 1975.

Roe, Frank Gilbert. *The North American Buffalo: A Critical Study of the Species in its Wild State*. 2d ed. Toronto: University of Toronto Press, 1970.

Rosenberg, Nathan. *Technology and American Economic Growth*. New York: Harper & Row, 1972.

Ruhl, Harry D. "Hunting the Whitetail." In *The Deer of North America: The White-tailed, Mule and Black-tailed Deer, Genus Odocoleus: Their History and Management*, edited by Walter P. Taylor. Harrisburg, Pa., and Washington, D. C.: The Stackpole Company and the Wildlife Management Institute, 1966.

Schmitt, Peter J. "Call of the Wild: The Arcadian Myth in Urban America." Ph.D. diss., University of Minnesota, 1956.

Schorger, A. W. "A Brief History of the Steel Trap and Its Use in North America." *Transactions of the Wisconsin Academy of Science*, Part 2, 140 (1951), 171-99.

———. *The Passenger Pigeon: Its Natural History and Extinction*. Madison: University of Wisconsin Press, 1955.

Schulz, William F., Jr. *Conservation Law and Administration. A Case Study of Law and Resource Use in Pennsylvania*. New York: Ronald Press, 1953.

Silver, Helenette. *A History of New Hampshire Game and Furbearers*. New Hampshire Fish and Game Department, Survey Report No. 6, May 1957.

Skinner, John E. *An Historical Review of the Fish and Wildlife Resources of the San Francisco Bay Area*. Water Projects Branch Report #1, The Resource Agency of the California Department of Fish and Game. Sacramento, June 1962.

Slobodkin, Lawrence. *Growth and Regulation of Animal Populations*. New York: Holt, Rinehart and Winston, 1961.

Smallwood, William Martin. *Natural History and the American Mind*. New York: Columbia University Press, 1941.

Smith, V. Kerry, ed. *Scarcity and Growth Reconsidered*. Baltimore: Johns Hopkins University Press, for Resources for the Future, 1979.

Smith, Vernon L. "Economics of Production from Natural Resources." *American Economic Review* 63 (June 1968), 409-31.

———. "On Models of Commercial Fishing." *Journal of Political Economy* 77 (March/April 1969), 181-98.

Stahr, Elvis J., and Charles H. Callison. "The Role of Private Organizations." In Council on Environmental Quality, *Wildlife and America*, pp. 498-511.

Stene, Edward O. *The Development of Wildlife Conservation Policies in Kansas*. Governmental Research Series #3. Bureau of Government Research. University of Kansas, 1946.

Swanson, Evadine Burris. "The Use and Conservation of Minnesota Game, 1850-1900." Ph.D. diss., University of Minnesota, 1940.

Tillett, Paul. *Doe Day: The Anterless Deer Controversy in New Jersey*. New Brunswick: Rutgers University Press, 1963.

Tober, James. "The Allocation of Wildlife Resources in the United States, 1850-1900." Ph.D. diss., Yale University, 1973.

Trefethen, James. *Crusade for Wildlife: Highlights in Conservation Progress*. Harrisburg, Pa. and New York: The Stackpole Company and the Boone and Crockett Club, 1961.

———. "Wildlife Regulation and Restoration." In *Origins of American Conservation*, edited by Henry Clepper for the Natural Resources Council of America. New York: The Ronald Press Co., 1966.

———. *An American Crusade for Wildlife*. New York: Winchester Press and The Boone and Crockett Club, 1975.

Van Brocklin, Ralph M. "The Movement for the Conservation of Natural Resources in the United States before 1901." Ph.D. diss., University of Michigan, 1952.

Van Wagenen, Jared, Jr. *The Golden Age of Homespun*. New York: Hill and Wang, 1953.

Wagner, Frederic H. "Ecosystem Concepts in Fish and Game Management." In *The Ecosystem Concept in Natural Resource Management*, edited by George M. Van Dyne. New York: Academic Press, 1969.

Wertenbaker, Thomas Jefferson. *The Puritan Oligarchy: The Founding of American Civilization*. New York: Charles Scribner's Sons, 1947.

Wilson, Edward O., and William H. Bossert. *A Primer of Population Biology*. Sunderland, Mass. Sinauer Associates, 1971.

Young, Stanley Paul. *The Wolf in North American History*. Caldwell, Idaho: The Caxton Printers, 1946.

———. "The Deer, the Indians, and the American Pioneers." In *The Deer of North America*, edited by Walter P. Taylor. Harrisburg, Pa., and Washington, D.C.: The Stackpole Company and the Wildlife Management Institute, 1956.

Ziswiler, Vinzenz. *Extinct and Vanishing Animals*. Rev. English ed. by Fred and Pille Bunnell. New York: Springer-Verlag, 1967.

INDEX

ABOUT THE AUTHOR

James A. Tober is a member of the faculty at Marlboro College, Marlboro, Vermont, and a Research Affiliate in the Program on Non-Profit Organizations of the Institution for Social and Policy Studies at Yale University.